Developing Markets for Agrobiodiversity

Securing Livelihoods in Dryland Areas

T0347466

Developing Markets for Agrobiodiversity

Securing Livelihoods in Dryland Areas

Alessandra Giuliani
Bioversity International

First published by Earthscan in the UK and USA in 2007

For a full list of publications please contact:
Earthscan
2 Park Square, Milton Park, Abingdon, Oxfordshire OX14 4RN
711 Third Avenue, New York, NY 10017

First issued in paperback 2014

Earthscan is an imprint of the Taylor & Francis Group, an informa business

Notices

Practitioners and researchers must always rely on their own experience and knowledge in evaluating and using any information, methods, compounds, or experiments described herein. In using such information or methods they should be mindful of their own safety and the safety of others, including parties for whom they have a professional responsibility.

Product or corporate names may be trademarks or registered trademarks, and are used only for identification and explanation without intent to infringe.

The views in this book are the author's own and do not necessarily represent those of Bioversity International.

ISBN 13: 978-1-84407-468-6 (hbk)
ISBN 13: 978-1-138-00203-6 (pbk)

Typesetting by Bioversity International, Rome, Italy
Cover design by Susanne Harris

A catalogue record for this book is available from the British Library

Library of Congress Cataloging-in-Publication Data

Giuliani, Alessandra.
    Developing markets for agrobiodiversity : securing livelihoods in dryland areas /Alessandra Giuliani.
        p. cm.
    Includes bibliographical references.
    ISBN-13: 978-1-84407-468-6 (hardback)
    ISBN-10: 1-84407-468-4 (hardback)
    1. Farm produce—Syria—Marketing—Case studies. 2. Agrobiodiversity—Syria—Case studies. I. Title.
    HD9018.S95G58 2007
    381'.41095691—dc22

                                                                    2007008616

Contents

List of boxes, figures and tables

BOXES

FIGURES

TABLES

Foreword

Bioversity International (formerly IPGRI) is implementing a new strategy: *Diversity for Well-being*. At the heart of the strategy is our commitment to demonstrate and increase the benefits of using biodiversity for local communities and to meet the Millennium Development Goals of reducing poverty, improving health and nutrition, and conserving natural resources.

A concrete way we are demonstrating and adding value to plant genetic resources is through markets. Many products and uses of biodiversity have potential to enter markets and provide healthy products for consumers and increase the incomes of rural households.

This study in Syria uses actual case studies of how communities have developed markets for local biodiversity-derived products. The data and the processes documented in this study show the potential of biodiversity, including otherwise neglected and underutilized species, to make a significant contribution to livelihood security in communities that live in difficult environments with unique resources. The study also highlights the importance of local cultural knowledge and institutions in sustainable development of markets for biodiversity products.

Bioversity thanks the many people and organizations that contributed to this work, and will continue to work with communities to help them market and benefit from biodiversity.

E. Frison
Director General

Acknowledgements

I would first of all like to thank all the people who provided information and shared their time, resources and knowledge. Without their friendly collaboration and input, this report could have not been written. So, sincere thanks go to the community members interviewed in Syria, farmers and traders, forestry guards, representatives of the Ministry of Agriculture and Agrarian Reform, and professors from the universities of Aleppo and Damascus.

The support of my colleagues Pablo Eyzaguirre and Melinda Smale has been invaluable in guiding the work and the analysis of the data, structuring the report and revising the texts. Thanks to Eric Van Dusen for his advice on methods and tools.

Thanks also to colleagues in Bioversity's Central and West Asia and North Africa (CWANA) office (in particular Stefano Padulosi and George Ayad, for guiding my work planning, and Professor Adnan Hadjhassan, Firas Shuman, Amer Ibrahim Basha, Silvia Bolognini and Laura Ventura), who together with the survey assistants (Bashar, Rasha, Robin, Yamam and Emad) and Nashwa Abdulkarim from UNDP, greatly contributed to the field work. Thanks to Markus Buerli for his constant, useful critical comments and support.

The work reported here was carried out by the author, Alessandra Giuliani, as part of a field research study during her assignment at Bioversity CWANA (under the programme sponsored by the Italian Ministry of Foreign Affairs).

Great thanks to Judith Thompson for revising and improving the text.

Thanks to Thorgeir Lawrence and Paul Neate for the editing and typesetting on behalf of Bioversity, and to Patrizia Tazza and Frances Ferraiuolo for graphics and layout.

All the photographs were taken by the author (© Alessandra Giuliani 2007), and, unless otherwise indicated, were taken during the fieldwork in Syria.

Abbreviations

CBD	Convention on Biological Diversity
CBS	Central Bureau of Statistics [Syria]
CGIAR	Consultative Group on International Agricultural Research
CIP	Centro Internacional de la Papa / International Potato Center
CIRAD	Centre de coopération internationale en recherche agronomique pour le développement [France]
CNR	Consiglio Nazionale della Ricerca [National Research Council, Italy]
DFID	Department for International Development [UK]
FAO	Food and Agriculture Organization of the United Nations
GDP	gross domestic product
GEF	Global Environmental Facility
GFAR	Global Forum on Agricultural Research
GFU	Global Facilitation Unit for Underutilized Species
GTZ	Agency for Technical Cooperation [Germany]
ICARDA	International Center for Agricultural Research in the Dry Areas
ICUC	International Centre for Underutilised Crops
IDRC	International Development and Research Centre [Canada]
IFAD	International Fund for Agricultural Development
INRA	Institut national de la recherche agronomique [France]
InWEnt	Capacity Building International
IRD	Institut de recherche pour le développement [France]
LS	Syrian Pounds (US$ 1 ≈ LS 51.5 at the time of writing)
MAAR	Ministry of Agriculture and Agrarian Reform [Syria]
MACAB	Marketing Approach to Conserve Agricultural Biodiversity

MAP	medicinal and aromatic plant
MDG	Millennium Development Goal
NGO	non-governmental organization
NTFP	non-timber forest product
NUS	neglected and underutilized species
PGRFA	Plant Genetic Resources for Food and Agriculture
PMCA	Participatory Market Chain Approach
R&D	research and development
SINGER	[CGIAR] System-wide Information Network for Genetic Resources
UNCTAD	United Nations Conference on Trade and Development
UNDP	United Nations Development Programme
UNEP	United Nations Environment Programme
WCED	World Commission on Environment and Development

1

Introduction

Neglected and underutilized plant species (NUS) are often called 'minor' because in global production and market value terms they are of minor importance when compared with major staple crops and other agricultural commodities. Nevertheless, they are sources of food, herbal remedies and income for poor rural communities. Analysis of the socio-economic and market aspects of such NUS is one of the research areas that can foster their sustainable conservation and use.

With the objective of contributing to a better understanding of livelihood options and deployment of NUS, this pilot study was conducted in Syria on some selected wild and cultivated NUS from dryland agro-biodiversity environments. The study examined the market organization and the characteristics of the actors involved, and identified initial, local-level constraints on generating greater market value for these species. The study involved an initial value-chain analysis to look at the production-to-consumption chain, the organization of the production system and the roles of the actors involved in the collection, cultivation, processing, value-adding and trading of the products; and followed a livelihoods approach to look at the impact on the livelihoods of rural communities of the activities related to these plant species.

Data confirm the strategic role played by these traditional and underutilized species in the livelihood strategies of rural households in the region, and the importance of strengthening markets if these strategies are to be supported and the species conserved *in situ*. However, interventions are needed at different levels to increase market potential, including capacity building and knowledge sharing on cultivation practices and marketing strategies, reorganization of the market relationships, and support for private sector partnerships. Generating market value would serve the dual goals of supporting rural livelihoods and promoting biodiversity maintenance.

The information presented here is addressed to various stakeholders, such as policy-makers, extension workers, scientists, students, research and development (R&D) organizations and non-governmental organizations (NGOs), with the hope that they may find some useful facts that encourage further research and studies, and, consequently, development actions to improve the market for these species, benefiting the poor, chain actors. The information is also addressed to the producers, processors and traders of these species in Syria, to foster their interest in their activities and to provide a record of their valuable traditional knowledge.

2

Study context: the potential of neglected and underutilized species to contribute to livelihoods

MOTIVATION FOR THE STUDY AND RESEARCH OBJECTIVES

Research has showed that NUS play a fundamental role in the livelihoods of rural poor communities, representing an available source of income, nutrition and medicinal remedies manageable using traditional knowledge. However, these species have unexploited potentials.

Research to enhance the use of NUS is needed in the field of biology and agronomy. However, to find ways to foster the sustainable conservation of natural resources for the benefit of poor communities relying on them, investigations of markets are needed, including social, cultural, economic and policy aspects. Combined investigations can help assess how to promote a better use of these species and how the market could be organized to allow a more equitable distribution of income along the market chain.

The aim of this research was to contribute to a better understanding of livelihood options and deployment of NUS, using concrete case studies focusing on a number of selected species important in the local agro-biodiversity and production systems. The study objectives were to explore the role of some NUS in the livelihoods of rural communities in Syria and to look at ways of generating market value from the products derived from these species to serve the dual goals of not only supporting rural livelihoods, but also biodiversity maintenance of the genetic diversity through their use. The assumption to be tested was that NUS were available and easily accessible resources for poor rural communities, and that through marketing their derived products, those NUS could contribute substantially to generation of household income. Maintaining the genetic resources of the species would lead to sustainable income-generation opportunities.

The research is not meant to be a market analysis of the products from NUS. The focus is on the nature and the degree of the involvement of the people engaged in the market chain, and on the driving forces that keep stakeholders active in the sector. What is analyzed is the process through which biodiversity products derived from NUS enter the market and how stakeholders use these species to contribute to their livelihoods, as well as to assess what interventions are needed at the various levels to support rural livelihoods while at the same time conserving biodiversity.

The study outputs were outlines of the market chains and the actors involved for the selected species; characterization of the livelihood assets of the actors; a profile of their social, demographic and economic characteristics; identification of local-level constraints limiting the full deployment of the selected NUS and the generation of greater market value from the species; and the identification of the needs and opportunities for boosting their uses to support development of rural community livelihoods.

The investigation looked at the market chain and at the needs and opportunities faced by each actor involved in it. The results will help identify where further studies are needed to develop support interventions along the chain and actor levels to channel benefits to low-income households that utilize important stocks of biodiversity.

The research was implemented in two steps by applying, first, market-chain methods, and, second, a livelihoods assets survey related to the selected NUS and their users, such as producers in rural communities and traders in some urban districts in various areas in Syria.

The data and information collected for this research was gathered in the period June 2003 to December 2004, with analysis extending into late 2005.

WHAT ARE NUS?

NUS are often called 'minor' because of their small total economic value in commercial production and trade compared with staple crops and agricultural commodities.

A wide range of terms are used for underutilized plant species, including 'minor', 'neglected', 'local', 'traditional', 'underexploited', 'underdeveloped', 'orphan', 'lost', 'new', 'niche', 'promising 'and 'alternative' (Padulosi et al. 2002).

Bioversity International (formerly IPGRI) defines these species as neglected and underutilized plant species, not crops, since wild, managed and cultivated species are taken into consideration. These plant species may belong to any category, from fruit and nut trees to leafy vegetables, from functional herbs (or medicinal and aromatic plants [MAPs]) to cereals, from legumes to forest trees, from forages to roots and tubers. Bioversity classifies these species as *underutilized* since in the past they were grown more widely or intensively but are now falling into disuse for a variety of agronomic, genetic, economic and cultural reasons. Some species may be widely distributed around the world but tend to occupy special niches in the local ecology and in local production and consumption systems. Farmers and consumers are using these species less because they are not competitive with other species in the same agricultural environment. However, NUS are well adapted to particular agro-ecological niches and marginal areas, and are cultivated and utilized using indigenous knowledge. These species are also characterized by fragile or non-existent seed supply systems (Padulosi and Hoeschle-Zeledon 2004). The decline of these species may erode the genetic base and prevent distinctive and valuable traits being used in crop adaptation and improvement (Eyzaguirre et al. 1999).

Bioversity defines *neglected* species as those that are grown primarily in their centres of origin or centres of diversity by traditional farmers, where they are still important for the subsistence of local communities. Although these species continue to be maintained by sociocultural preferences and the ways in which they are used, they remain inadequately documented and neglected by formal research and conservation (Eyzaguirre et al. 1999). Hence, these species are called *neglected* as they receive scarce attention from national agricultural and biodiversity conservation policies, are almost ignored by scientific research and development and are barely represented in *ex situ* gene banks (Padulosi and Hoeschle-Zeledon 2004). The sheer number of species is a challenge to botanists and agricultural researchers. Very limited information about these plant species may be available in libraries and databases around the world, but extensive information nevertheless does exist.

The *New Agriculturist* (2004) reports that minor crops may be described as 'neglected' or 'underutilized', but in their native areas, where people depend on them as important components of subsistence farming systems, these crops are neither. Moreover, today's small-scale farmers have not only their own life experience, but also hold the accumulated knowledge of previous generations. However, capturing this indigenous knowledge before it is lost forever, and indeed so that it can be more widely used and benefit other communities, is a significant challenge (*New Agriculturist* 2004).

It is important to emphasize that the definitions of underutilized and neglected are still under discussion. The debate is still open on what these species are and how the definition could be quantifiable. Underutilized is commonly applied to refer to species whose potential has not been fully realized. However, Padulosi et al. (2002) write that the term gives no information about geographical, social or economic implications. With regard to *geographical distribution*, often a species could be underutilized in some regions but not in others. Regarding *social* and *economic implications*, many species represent an important component of the daily diet of millions of people, but their poor marketing conditions make them largely underutilized in economic terms. With regard to the *time factor*, the degree of underutilization of a plant species may be subject to a sudden improvement due to dynamic marketing systems present in some countries, while the same crop may continue to be poorly marketed and managed by researchers in others (Padulosi and Hoeschle-Zeledon 2004).

There have been efforts to identify criteria and categories that characterize NUS. The Global Facilitation Unit for Underutilized Species (GFU) identified 11 criteria that characterize NUS (Table 1).

Table 1. *Criteria characterizing NUS.*

Require only limited external inputs for production
Suitable for organic production
Suitable for cultivation on marginal land (poor soil fertility, etc.)
Suitable for stabilization of fragile ecosystems
Fit into small-scale farming systems
Possess traditional, local and/or regional importance
Easy to store and process by resource-poor communities
Market opportunities available
Possess high nutritional and/or medicinal value
Offer multiple uses
Traditional knowledge

Source: GFU, 2002.

Similar lists of criteria were developed by De Groot and Haq (1995), and Von Maydell (1989) identified conditions characterizing NUS for his study developed for semi-arid regions.

NUS include non-timber forest products (NTFPs), which are plant species that can be found wild in forest areas and rangelands, which grow spontaneously and are harvested by local communities for various uses, but not exploited for wood.

GLOBAL INITIATIVES ON NUS

New research on NUS, including socio-economic studies, was also prompted by the global attention that helped raise awareness concerning biodiversity. The gradual global increase in attention towards NUS was initiated by the awareness-raising process initiated by the United Nations Environment Programme (UNEP) Convention on Biological Diversity (CBD 1992) and then the 1996 Global Plan of Action for the Conservation and Sustainable Utilization of Plant Genetic Resources for Food and Agriculture (PGRFA).

Section 12 of the Global Plan of Action fosters the *Promotion of development and commercialization of under-utilized crops and species* (FAO 1996) with a:

- long-term objective of contributing to agricultural diversification, increasing food security and improving farmers' livelihoods, to promote the conservation and sustainable management of under-utilized species and their genetic resources; and an
- intermediate objective of developing appropriate conservation strategies and sustainable management practices for under-utilized species; improving selected species; [and] improving the marketing of under-utilized crops.

The second activity is closely linked to on-farm management of plant genetic resources and development of new markets for local varieties, found in Section 14 on *Developing new markets for local varieties and "diversity rich" products*, with the following objectives:

- *long-term*: to stimulate stronger demand and more reliable market mechanisms for local agricultural products;
- *intermediate*: to encourage farm suppliers, food processors and distributors and retail outlets to support the creation of niche markets for diverse products.

In Article 6 of the 2001 International Treaty on PGRFA on Sustainable Use of PGRFA, paragraph E is on policy and legal measures 'promoting, as appropriate, the expanded use of local and locally adapted crops, varieties and under-utilized species'.

The Consultative Group on International Agricultural Research (CGIAR), the alliance of international agricultural research centres, is currently focusing on ecosystems approaches and is addressing poverty in marginal areas with specific attention to neglected and underutilized species, as they are a crucial resource in the livelihoods of communities in remote areas. These species are considered a source for food security and poverty alleviation (MSSRF 1999).

In 2000, underutilized crops were given high priority in the commodity chain agenda by the Global Forum for Agricultural Research (GFAR), where their role in raising income of the rural poor was emphasized. Later, in 2002, a Global Facilitation Unit (GFU) under the umbrella of GFAR was established with funding from the German Ministry of Economic Co-operation and Development (BMZ) to promote the use of underutilized crops. In 2003, GFU, in a joint effort with the German Agency for Technical Cooperation (GTZ) and Capacity Building International (InWEnt), organized an International Workshop on Underutilized Species, with the aims of: identifying strategic elements for the promotion and sustainable utilization of underutilized species; and identifying potential actors for implementation (Guendel et al. 2003).

The role of NUS in global food security, effective nutrition and income generation was highlighted in the International Consultation held in Chennai, India, in April 2005 to examine the value of agricultural biodiversity in meeting the Millennium Development Goals (MDGs) (IPGRI/GFU/MSSRF 2005).

CURRENT STATE OF KNOWLEDGE ON THE ROLE OF NUS IN LIVELIHOODS

Increased reliance on major food crops has been accompanied by a shrinking of the food basket that humankind has relied upon for generations (Prescott-Allen and Prescott-Allen 1990). The narrowing base of global food security is limiting livelihood options for the rural poor, particularly in marginal areas. The shrinking of agricultural biodiversity has reduced both the intra- and interspecific diversity of crops, increasing the level of vulnerability among users, particularly the poorer sections. For some, diversity in crops is a necessity for survival rather than a choice.

While modern crop production predominantly involves only hundreds of the many thousands of the known food plants globally, ethno-botanic surveys indicate that thousands of traditional species are still to be found wild or cultivated in many countries (IPGRI 2002), yet they are largely ignored by scientific research and many may be under threat. Such species are still managed by poor communities in remote areas, and the potential value remains underexploited. These species, important in local farming systems, have great potential in a future sustainable global food system (*New Agriculturist* 2004). These minor species have a comparative advantage in marginal lands, where they have been developed to withstand stress conditions and contribute to sustainable production with low-cost inputs. They have a strategic role in fragile ecosystems, such those of arid and semi-arid lands, highlands, steppes and tropical forests, where they have a role to play in sustained use of the environment and the restoration of degraded lands (De Groot and Haq 1995).

Global studies report that rural communities—often regarded as poor on a global scale—depend on intra- and interspecies diversity to cope with climatic risk, to meet phytopathological challenge, to match them to specific soil and water regimes, and to meet a range of consumption needs when markets are unreliable (Engels et al. 2002). Diversification in agricultural systems is indeed an important asset for those fragile social groups who may never be able to afford certain commodities. For these reasons, securing the resource base of NUS, particularly in developing countries, is crucial to maintain the 'safety net' of diversified food and natural products that has provided options to address food needs in a sustainable way (Eyzaguirre et al. 1999).

NUS are part of the biological assets, often freely available, of the rural poor and provide a variety of products and benefits to them, such as foods, medicinal products, material for handicrafts, and additional income. Moreover, these plants are often related to the cultural and religious values of the local communities, hence preserving those plants helps preserve the cultural identity of those communities (IPGRI 2002). They add diversity to the diet and, because they are part of the traditional food system, they are readily accepted (Johns and Eyzaguirre 2002). Artistic, landscape and cultural values of NUS in industrialized countries are also recognized (Monti 1997).

Nevertheless, the undervaluing of the economic value of NUS used by rural people also contributes to loss of crop genetic diversity and erosion and simplification of ecosystems, thus restricting options that might benefit future generations of rural people and urban consumers who would be willing to pay for these products of biodiversity, thus limiting species' private and public diversity values, where private value is measured in terms of objectives the farmer pursues for their own personal benefit. The public value of genetic diversity refers to the welfare of society rather than of its individuals (value for future generations, for potential disasters and unforeseen events) (Smale and Bellon 1999).

The present research seeks to test the hypothesis that NUS have unexploited potentials to play a fundamental role in the livelihoods of rural communities, by the production and marketing of NUS, using traditional and available technologies. The market for these species should be maintained and enhanced in order to enhance the livelihood opportunities of the people involved in the activities, and at the same time promoting conservation of the genetic resources through their sustainable use.

Caper collectors in Jabal-al-Hoss, Aleppo Province, Syria.

The published results from a range of recent studies confirm the hypothesis of unique benefits of NUS for the livelihood of poor rural communities and the associated ecosystems. Some—among many others—of this research is reported to show the increased interest in the role of these species. However, much has still to be done, and more formal and coordinated studies to further prove this hypothesis should be organized and conducted.

In Africa, a study carried out by McClintock (2004) in Senegal and Mali on Roselle (*Hibiscus sabdariffa*) shows the importance that the cultivated species of Roselle has among the rural communities, being a multi-functional plant with ecological, dietary, medicinal and income benefits, in particular for women. Various studies on traditional leafy vegetables in Africa (Chweya and Eyzaguirre 1999) have documented their importance for providing affordable and nutritious vegetables for local consumption, generating income through trading while maintaining the stability of ecosystems. In West African countries, Centre de coopération internationale en recherche agronomique pour le développement (CIRAD) and national research institutes are carrying out studies to improve grain processing techniques for small-level enterprises and women's groups for Fonio (*Digitaria exilis*), a plant supplying food to several million people during the most difficult months of the year (the 'hungry gap') when other food resources are scarce. The International Centre for Underutilised Crops (ICUC) was involved in a farmer's participatory survey in South Africa and assessed the rich traditional knowledge of the nutritional properties and adaptation to local environment of Livingstone potato (*Plectranthus esculenta*), cucurbits and *Amaranthus* spp., among other species kept by the women of rural communities. The experience of the South and East Africa Network for Underutilized Crops (SEANUC) shows that women continue to use NUS and are interested in crop diversification activities, including community participatory production and community-level processing and marketing, to enhance food security and contribute to sustainable rural livelihoods.

In Asia, food derived from Finger millet (*Eleusine coracana*) in India, for example, gave reduced risk of heart disease, while Bitter gourd (*Momordica charantia*) and Fenugreek (*Trigonella foenum-graecum*) contain compounds that directly improve the body's ability to respond to insulin (Johns and Sthapit 2004). In Viet Nam, studies showed that rural communities use some wild-harvested NTFPs for food, income, farm inputs and cultural and religious functions. Even in isolated areas, the sale of those products provides a substantial share of the income of local households (FAO 1995).

In the Andean countries, as part of a global initiative supported by IFAD and Bioversity aimed at increasing the role of NUS in food security and income generation among poor rural communities, several participatory studies have shown the economic and social value of Andean pseudograins (Quinua or Quinoa [*Chenopodium quinoa*] and Canahua or Cañihua [*C. pallidicaule*]) among the households living in rural areas.

In Pacific countries, on-farm conservation studies have been conducted and funded by Institut de recherche pour le developpement (IRD) and CIRAD on the impact of cultivating different varieties of Taro (*Colocasia esculenta*) on small farms (Caillon and Lanouguère-Bruneau 2003).

Activities in the field of marketing agricultural products

Many organizations that have focused their work on the improvement of agricultural production and productivity are now looking also at product processing and marketing (Hellin and Higman 2005). The CGIAR's strategic priorities recognize the importance of markets and market access for poverty reduction. Improvements in production and productivity must be linked to increased access and efficiencies in markets in order to improve the livelihoods of the poor (CGIAR 2004). This is in line with the major attempt to accomplish the Millennium Development Goals (rural poverty reduction in particular).

Bioversity's global mandate on NUS includes the study of the market chain to identify potential improvement in value-addition and marketing aspects along the chain.

Understanding which markets are currently and potentially most important for the poor is critical in order to plan both market and non-market livelihood interventions. In terms of approaches, the analysis of livelihood assets and livelihood strategies of small-scale producers, processors and traders is a starting point in ensuring that markets benefit the poor. To address the market and its roles in poverty reduction—a dimension not overly emphasized by the livelihood approaches (Dorward et al. 2002)—R&D organizations also apply market-chain methods. In market-chain analysis, some of the challenges that need to be addressed in order to benefit the poor are improving small-scale farmer competitiveness and farmers' organizations (Biénabe and Sautier 2005); institutional capacity building, in particular in terms of access to information (Kydd 2002); and the reinforcement of links and trust among actors in the market chain (Best et al. 2005).

Value-chain analysis has recently been applied to the understanding of commodity chains in developing countries, with integrated and participatory approaches (Lundy et al. 2004). A few studies have been carried out on the link between biodiversity and market and the impact on livelihoods, developing and applying a participatory market chain approach, (Bernet et al. 2005). A market analysis and development field manual was developed by FAO (2000) to provide a framework for planning forest product enterprises with local communities.

OUTLINE OF THE REPORT

This document reports the result of the research and study analysis that aimed to evaluate some assumptions regarding the potential of NUS to enhance the livelihoods of the people involved in the market activities of their derived products. The report is structured in six sections. Following this *Study context*, that introduces the background against which the research was carried out, providing definitions, setting and objectives of the study, the second section, *Methods* explains the methodologies used in this pilot study. Section three, *Species and their uses,* describes the target species and their uses in the country. Sections four—*Findings: Species and Market Chain*—and five—*Findings: species and livelihoods of chain actors*—describe the findings emerging from focus-group interviews and household surveys, respectively, concerning the market chain and the household livelihoods of those involved. The last section presents *Conclusions and recommendations.*

3
Methods

SELECTION OF THE TARGET SPECIES

Choosing the right species to focus on from a broad group of potential candidates was a necessary step in making the best use of the limited resources available for this study.

Six species were selected, in consultation with local users of the species, farmers, researchers and academics, from a list of species characterized as NUS according to GFU criteria (Table 1) for dryland agro-biodiversity.

The species selected were Jujube, Fig, Caper, Laurel, Purslane and Mallow. To allow comparisons, the selection included different species categories (fruit trees; medicinal and aromatic plants; and vegetables) and cultivated versus wild species (Table 2).

Table 2. *Target species.*

Category	English name	Latin name	Family	Cultivated or wild
Fruit trees	Jujube	*Ziziphus jujuba* Miller	Rhamnaceae	Mainly wild
	Fig	*Ficus carica* L.	Moraceae	Mainly cultivated
Medicinal and aromatic plants	Caper	*Capparis spinosa* L.	Capparidaceae	Only wild
	Laurel	*Laurus nobilis* L.	Lauraceae	Mainly wild
Vegetables	Purslane	*Portulaca oleracea* L.	Portulacacee	Mainly cultivated
	Mallow	*Malva silvestris* L.	Malvaceae	Mainly wild

In the selection process, the extent (limited) and quality (poor) of information available on these species was considered, with the exception of Fig, where statistics (MAAR 2003) on production and processing existed.

Almost all the selected species had market potential for poor rural Syrian communities that collected or cultivated them. The selected species and their related products have a consumer demand in the local market and—for some products—also in international markets. The targeted species are very adaptable to local site conditions and they are compatible with different land uses, these species being a source of secondary income. They generally require low financial and technical inputs.

ANALYSIS OF THE MARKET CHAIN

Market-chain analysis was used in the present study to investigate how market relationships affect the livelihoods of people involved in the chain, in particular vulnerable people. Vulnerable people may be involved at the production level either as small-scale farmers, as collectors of wild species from public lands, or as casual, seasonal or permanent labour. The market-chain analysis was also used to identify the people involved to whom to address a livelihood survey, and to foster greater understanding of sustainable livelihoods, in terms of interventions affecting the assets, policies and institutions of the Sustainable Livelihoods Framework, at different level of the market chain. Among the various market-chain methodologies, value-chain analysis and *filière* methodology were used as a reference for the market-chain study of the present research. These two methodologies—value-chain analysis and filière—are considered in the next sections.

Value-chain analysis

Value-chain methodology developed from two categories of literature: business literature on strategy and organization (Porter 1990) and the literature of *global commodity chains*, promoted by Gereffi (1994) and developed in numerous studies in the late 1990s (UNCTAD 2000). The value chain is defined as 'the full range of activities which are required to bring a product or service from conception, through the intermediary phases of production, delivery to final consumers, and final disposal after use' (Kaplinsky and Morris 1999). Such analysis focuses on the interaction of actors at each step of the production system (from raw producer to consumer), as well as the linkages within each set of actors. Blowfield (2001) affirms that the value chain is one way in which the implication of market linkages for poor people may be explored, and that the market can be integrated into the Sustainable Livelihoods Framework in the Policies, Institutions and Processes (PIP) box (Annex 1). Value-chain analysis involves exploring the complexity of the actors involved as it affects the production to consumption process. It incorporates production activities (cultivation, manufacturing and processing), non-production activities (design, finance, marketing and retailing), and governance.

For this study, some methodological aspects of value-chain research, described by Kaplinsky and Morris (2001), have been utilized and adapted. The selected point of entry for the research analysis defines which links and which activities in the chain are to be the subjects of special enquiry. For this research, the starting point for the value-chain study was elaborated as follows:

- *Primary area of research interest:* the issues of responsible management of biodiversity and the support of rural livelihood.
- *Entry points:* the stakeholders from rural communities in Syria, involved in activities related to the selected species.
- *What to map:* activity organization of the whole chain, from collectors and growers to retailers, and the distribution of values added along the chain.

Having identified the value chain in question, a conceptual 'map' of the 'tree' of input-output relationships was constructed. The map included some of the following items coming from the data collection:

- the range of activities in the chain;
- the various actors performing the activities;
- the physical flow of the product along the chain and links among the actors;
- the net margin for the chain actors;
- the destination of sales, such as to wholesalers or retailers; and
- exports, and to which region.

The essential features of quality of the product are reported. Some aspects of the market environment have been taken into consideration, such as market organization, definition of the price, and the impact of policies and legal issues on the whole chain. Furthermore, constraints and opportunities for the different market-chain actors were identified. Two types of value chains have been identified in the literature: producer-driven and buyer-driven chains (Gereffi 1994). In the case of agriculture, there are both types of market structures, though increased consolidation in the retail sector has led to an increase in the power of retailers in food distribution. Also, in the case of NUS, these being agricultural or horticultural products or derived products, the chains are usually buyer-driven. In the recommendations, upgrading issues from both the research and development point of view have been emphasized. In the context of value-chain analysis, upgrading takes the form of either developing new, higher-value market niches, or expanding the range of activities (e.g. a manufacturer expanding their activity into distribution or R&D). The role of the market environment is important in how such upgrading occurs, as is the support of government and other institutions (UNCTAD 2000). A complementary issue is the means by which benefits are distributed within the chain. This refers to the amount of benefit obtained by various actors in the chain, as well as ways in which actors try to improve their position within the chain. The management of the market chain was therefore also analyzed.

Filière methodology

In this study, definitions related to the filière typologies, developed by researchers at the Institute national de la recherche agronomique (INRA), France, and

CIRAD in the 1960s were reviewed. The notion of filière is used to describe studies where a given product is followed along a 'chain' or 'system' of activities from producer to the final consumer. The filière approach, similar to the value chain, describes the flow of physical inputs and services in the production of a final product. Filière studies dealt initially with local production systems and consumption, while areas such as international trade and processing were largely overlooked until the 1980s. Hugon (cited in Griffon 2002) conceives the agricultural and agro-food market as a complex of filière typology. He describes the typologies of filière as having four dominant modes (domestic, market, state and agro-business) with different characteristics in terms of: production system; mode of circulation; mode of utilization; place; time; stakeholders; coordination mode; and global function. In the real world, gradations and combinations of the four typologies can be observed. Many agricultural economies and market systems in developing countries, and in particular related to the use of traditional species, combine both domestic and market spheres.

LIVELIHOODS SURVEY OF THE CHAIN ACTORS

With market-chain analysis, four general categories of chain actors were identified:

- *collectors*: individuals or groups collecting the wild species from state and private lands;
- *growers*: small-scale farmers, cultivating and harvesting the selected species on their own land;
- *small-scale processors*: dedicated to transformation activities on the product; and
- *traders*: intermediaries (middlemen), wholesalers and retailers.

A livelihood survey assessed the livelihood capitals of households of the various categories of chain actors. The definitions used in this research to refer to livelihood assets are taken from the Sustainable Livelihoods Framework (DFID 1999) (Annex 1). According to the DFID framework, the Sustainable Livelihoods approach is a way of thinking about the livelihoods of poor people and how to improve performances in poverty reduction. The approach is presented under a Sustainable Livelihoods Framework, highlighting the main factors that affect people's livelihood, and typical relationships among these factors (the Sustainable Livelihoods Framework in Annex 1). Livelihood assets in the Sustainable Livelihoods framework was selected for its advantages, namely that it is both centred on people and aims to organize the various factors that constrain or provide opportunities and to show how these relate to each other; and its essence lies in recognizing the diverse dimensions of poverty, and the multiple strategies that poor people adopt to secure their livelihoods (Albu and Scott 2001). However, for research focusing on market functioning, the Sustainable Livelihoods Framework has some drawbacks. Though 'the market' is listed as part of the Policies, Institutions and Processes (PIP) box (Figure A1.1), in literature it is widely recognized that the framework does not give enough space to the analysis of the impact of the market (Ellis 2000; Hobley 2001). Dorward et al. (2002) argued that there is a lack of emphasis on markets and their roles in livelihood improvement and poverty alleviation. Failure to address the role of markets in livelihood analysis could result in a failure to identify livelihood opportunities and constraints arising from critical market processes, and institutional issues that are important for pro-poor market development.

In this research, the analysis of the value chain, combined with the livelihood strategy assessment of the chain actors, was a way to look at market aspects in the Sustainable Livelihoods study approach. Once the market-chain actors were identified through value-chain analysis, a household survey was carried out of these actors to assess their livelihoods capitals, and how they were influenced by the market activities related to NUS products that formed part of their livelihood strategy.

Furthermore, looking at the livelihoods assets (Figure 1) described in the Sustainable Livelihoods Framework, the importance of NUS for different livelihoods capital becomes clearer, in particular for human and social (for traditional characteristics and easy access) and natural capital (for high adaptability to the natural environment).

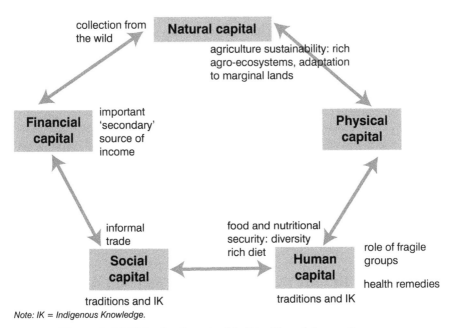

Note: IK = Indigenous Knowledge.

Figure 1. *NUS in the Sustainable Livelihood Assets Pentagon.*

DATA GATHERING TOOLS

Data gathering tools for market-chain analysis

For the market-chain analysis, checklists were used in focus-group interviews designed to identify the actors involved and to elicit information about the structure of the market chain and how activities were spread geographically within the country. Informal interviews were held with key informants, such as academics, researchers, a few traders of selected species, and policy-makers. The value-chain methodology guidelines were used and adapted for the development of checklists for informal interviews with the chain actors identified (including producers, processors, traders, policy-makers and market experts) for a number of sites in Syria, to collect information in a participatory manner. In collaboration with the actors involved, the organization of the activities along the chain, together with constraints to and opportunities for expanding the market, were identified. Based on the information gathered from the key informants interviewed and a number of chain actors interviewed, a map of product flow along the chain and the links among the actors could be drawn.

Data gathering tools in the household livelihood assets survey

For the household livelihood assets survey of chain actors, semi-structured questionnaires were used in interviews with individual market-chain actors to gather both qualitative and quantitative data about the livelihoods assets of those involved. Four different questionnaires were developed for the four categories: growers; collectors; processors; and traders. The respondents belonged to one of the categories in the market chain of the selected species, and were members of rural, forestry and peri-urban community households. The four questionnaires covered issues related to human, social, physical, financial and natural capitals (Annex 2). Income shares and returns to labour were investigated. Specific questions addressed the activities carried out by the actor interviewed, such as on cultivation and harvest, environment and collection, processing and trading, as well as on constraints and potentials. The aim of the survey was to highlight differences in the perception of the same issues or same species among the various agents along the market chain, as well as to identify at what level barriers and opportunities occurred. The unit of observation was the respondent, in most cases the head of the household, who was interviewed by a local survey assistant.

SITE SELECTION AND REPRESENTATION

The study was conducted in each of the five agricultural regions of Syria (NAPC 2003) (Annex 3). An effort was made to cover regions with differing topographical, social and economic characteristics. Criteria used to select ecosites included agro-ecological features, market importance, and presence of the selected species where they are cultivated or collected from the wild.

The five sites surveyed were in the north, the city of Aleppo and the agriculture pole of Idlib; in the west, coastal and forestry areas around the city of Lattakia and Tartus; in the centre, the cities of Hama and Homs; in the east, the desert region around the city of Raqqa and along the river Euphrates; and in the south, the region around Damascus and the agriculture pole of Sweida. Figure 2 shows the selected sites superimposed on the five Agricultural Settlement Zones of MAAR (2002) in Syria. These Agricultural Settlement Zones are relevant for various national policies in Syria and are based on annual rainfall levels. While Zone 1 gets the most precipitation, Zone 5 is desert (Annex 3).

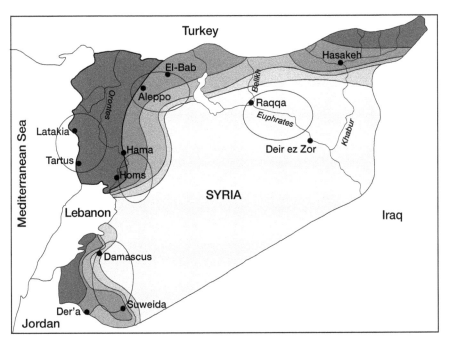

Figure 2. *The survey sites (ellipses) superimposed on a map of the five Syrian Agricultural Settlement Zones.*

The rural economy in Syria

This section gives an overview of the rural Syrian environmental context.

Agricultural production is a significant determinant of overall national economic growth. Agriculture has contributed a share of 27–32% to real GDP in the last decade (NAPC 2003). Accordingly, agriculture is by far the largest sector of the Syrian economy. The bulk of exports are agriculture based, manufacturing is based on agro-processing, a large share of trade and commerce is based on agriculture, and many services are linked to agricultural production (Sarris 2003). Moreover, a large proportion of employment derives from agriculture. In 2001, the agricultural sector provided employment to about 32% of the workforce. However, there is great variability in employment, due to varying climatic conditions, which have substantial impact in terms of seasonal labour force requirements in major crops (cotton and cereals) (CBS 2001). About 55% of the population lives in rural areas. The social surrounding affects agricultural development in many ways in Syria. At present, the social surrounding has a negative impact on agricultural development. Syrian agricultural development is affected in particular by the extent of illiteracy, a lack of ecological awareness, the system of land ownership and the system of rights to benefit from land use. Illiteracy affects agricultural development in various ways: illiterates accept only with difficulty modern technologies that seek to raise productivity and conserve nature; and illiterates often obstruct sound family planning, leading to high growth rates exceeding the support capacity of natural resources.

Regarding land ownership, in the interval between the agricultural censuses of 1981 and 1994, there was a considerable (26%) increase in the total number of holders (CBS 1999). Considering the fact that the total cultivable land did not changed much during this period, it can be concluded that there has been considerable fragmentation and subdivision of farms, despite laws and regulations that explicitly forbid it. This must have been the consequence of population growth coupled with long-standing social norms in Syria that require the subdivision of land among family members (Sarris 2001). This segmentation of agricultural land creates another constraint on rural development. The result of the inheritance system is small land plots that are not economically profitable, limiting possibilities for any scale economies in production and marketing.

Human capital is the dominant element or lynchpin in the relationship between the environment and development. Social customs, traditions and rules govern and shape the relationship among environment, agriculture and livelihoods. For rural peoples, these lie at the core of agricultural development. Recent UNDP/GEF studies (2000) indicate that poverty results from various elements, including fragmented and small-sized land-holdings, reduction in average rainfall, high cost of ground-water, deteriorating soil and poor agricultural marketing infrastructure. Poverty is thus mainly concentrated in the mountain regions and marginal and dry areas.

HOUSEHOLD SAMPLING SURVEY

Data on the population of growers, collectors, processors and traders of targeted species were not available. Consequently, it was not possible to select a random sample of growers, collectors, processors and traders of the target species at each site as a function of the respective population. So, members were selected from the population in some non-random manner. Surveyed units do not belong to statistically representative samples. The non-probability samples have been chosen using two purposive sampling methods for each site:

- *Snowball* or *chain sampling* was used for locating information-rich key informants (Patton 2002) from each category of respondents. The snowball approach consists of identifying someone who meets the criteria for being included in the study (for example, collector of species x at site y), and then to ask them to recommend others whom they know who also meet the criteria (Trochim 2001). This method is especially useful when you are trying to reach populations that are hard to find, as in the case of NUS.
- *Maximum variation sampling* was used across the various categories of stakeholders, species and sites to include a great deal of variation. This purpose-sampling strategy aims at capturing and describing the central themes that cut across a great deal of variation (Patton 2002), and identifying common patterns emerging from great variation by capturing core experiences among the same categories of stakeholders dealing with different species. Reflecting the different locations of the species at surveyed sites and the presence or absence of particular stakeholder categories, the breakdown of the distribution of the sample by species and market-chain actor is shown in Table 3.

Table 3. *Distribution of samples by species and market-chain actor.*

	Growers	Collectors	Processors	Traders	Total
Fig	36	0	19	24	79
Jujube	8	0	0	7	15
Caper	0	15	7	5	28
Laurel	3	3	2	15	23
Mallow	7	20	0	20	47
Purslane	19	15	0	26	60
Total	73	53	28	97	251

Without random samples, no rigorous statistical inferences can be made. However, the study highlighted some novel information that could encourage further studies and, consequently, development actions to improve the market for these species, thus benefiting the poor chain actors in these areas.

Survey implementation

Five young Syrian professionals participated in the household survey data collection. They were graduate and post-graduate students from the Universities of Aleppo, Lattakia and Damascus.

They were selected to gather information on their area of origin, so that, for instance, the survey assistant from the southern area carried out all the interviews in that area. This was intended to create an atmosphere of empathy and trust with the interviewees through familiarity of area habits, customs, dialects, etc. An introductory and instruction day was organized at the beginning of the survey to explain the aim of the research and the structure of the questionnaires. General guidelines on how to approach the respondents during the interviews were discussed and jointly developed. Guiding principles developed by FAO in the marketing extension guide *Market research for agroprocessors* (FAO 2003) for interviewing traders were followed. Assistants helped to test and adjust the questionnaires, and visits were paid to supervise and monitor their work at the five sites during the data collection period. Interviewee replies were reviewed together with the survey assistants, paying particular attention to misleading replies that might have been due to misunderstanding of the questions.

4

Species and their uses

The brief descriptions of the species given in this section is based on previous literature (published and grey), and on focus-group interviews with academics, researchers, farmers and traders.

Table 4 gives the English and Arabic names of the selected species.

Table 4. *English and Arabic name of the selected species.*

English name	Arabic name	Phonetic
Fig	تين	Tiin
Jujube	عناب	Annab
Laurel	غار	Ghar
Caper	قبار	Kabar
Purslane	بقلة	Bakleh
Mallow	خبيزة	Khebeseh

FIG

The Fig (*Ficus carica*, family Moraceae) probably originated in western Asia, and spread to the Mediterranean. Its geographical distribution matches that of the olive tree.

Today, the fig is a moderately important world crop. The typical fig-producing regions have mild winters and hot dry summers. The fig tree is tolerant of a wide range of environmental conditions, has a low chilling requirement, will withstand some frost and is tolerant of drought, although it grows most vigorously with abundant water. Figs can be grown on a wide range of soils, including heavy clays, loams and light sands, but ideally the soil should be well drained. The plant is moderately tolerant of high salinity (Tous and Ferguson 1996.). The fig plant is very adaptable, growing in all five areas surveyed, and it is believed to have been grown for more than 4000 years over a large area of Syria, despite agro-climatic variability and contrasting environments. Fig trees can be found in areas of less than 200 mm/yr of rain, and in areas with more than 1000 mm/yr, from sea level to more than 1500 m elevation, and in different soil types, as it tolerates a range of pH and salinity. It can grow in rocky areas and can withstand drought and high temperatures (ICARDA 2001). Figs are not usually seriously affected by pests.

Wild fig tree growing in the archeological site of Krak des Chevaliers, in Homs Province, Syria.

The area of figs in Syria has decreased significantly during the past three decades, from 21 623 ha recorded in 1970 to 10 740 ha in 2001. Much of this decline has been due to the replacement of figs by other fruit species that offer higher economic returns, such as olive and apple. However, production has fluctuated in recent years. From official data, there were about 2 723 000 fig trees in Syria in 2001, of which more than 80% were rain-fed. Collection from the wild trees was scattered, while production from cultivated trees, both irrigated and rain-fed was 40 019 t in 2001, of which about 30% was processed (MAAR 2003).

The value-chain analysis described in a later section focused only on the biggest production area in Syria: Kafranbel in Idleb Governorate (central survey site). Kafranbel is 735 m above sea level, with average rainfall of 460 mm/yr (i.e. in Zone 1 in the official typology). Fig cultivation in Kafranbel covers 3 678 ha, with 778 000 trees, more than 90% of which are rain-fed, and producing about 12 500 t in 2001 (MAAR 2003). There were about 2 000 households working on the production and natural processing of figs, equivalent to about 60% of the households in the area (Al Ibrahim 1997). In each household, an average of seven family members were involved in the fig production (parents and children from 7 years of age). There were no employees. Farmers generally dedicated 80% of their cultivation time to the production and processing of figs, and 20% to olive production (Al Ibrahim 1997). Cultivation of figs and olives provided one-third of these households' income. The period of the year dedicated to the production, harvesting and processing of figs was from February to November. The *baiadi* variety was mainly grown. Trees were productive after five years. This variety was considered to be the best one in the region for drying purposes, but not for fresh consumption. During the period 1990–99, fig production showed a significant increase (about 1000 t per year). In the same period, cultivated land decreased by about 567 ha/yr. The yearly household consumption of figs was about 40–50 kg for each family, corresponding to less than 5% of production, while the remaining 95% was commercialized. The fruit is usually consumed either fresh locally or in dried, canned or preserved form (jam or paste).

Local names and uses

Fig is called 'tiin' in Arabic (تين) everywhere in Syria. Many fig varieties are grown, with various local names that differ from area to area. In the Biological Diversity National Report, 33 fig varieties were listed (GEF/UNDP 2002). Some local names in Arabic are reported in Table 5, with names coming from their colour, shape or taste, their origin, the area where they are cultivated or the period when they are harvested. It is interesting to see the reference to *khan* in one local variety name. This refers to the resting places for travellers (with horses) along journey routes. These places became exchange places for goods, bought and sold by traders. There used to be a *khan* about every 40 km, and the name can be found in many villages in Syria (Issa, pers. comm.).

Table 5. *Local names of fig varieties in Syria.*

Local name in Arabic	Phonetic	Meaning
قرصاوي	Kersawi	Hitching
كفري	Kfairi	Wild
سويداوي	Swaiedi	From Sweida (Damascus Govenorate)
عربي سلطاني	Arabi sultani	Sultan Arab
أسود	Aswad	Black
بياضي	Baiadi	Fair (colour)
غريب	Ghraib	Foreign
برغلي	Burgely	Uneven (rough)
عسلي	Asali	Honey
فنري	Fenery	From the Fenery village (family name)
غزالي	Ghazaly	Gazelle
حبشي	Habashi	From Ethiopia
حلبي	Halabi	From Aleppo
حميري	Hmaery	Reddish
ادلبي	Idliby	From Idleb (Idleb Govenorate)
الخاني	Al Khani	From a khan – a resting place or commercial exchange point
خضيري	Khdaery	Greenish
ملكي	Malaki	Kingdom
صفراوي	Safraouy	Yellowish
سمكي	Samaky	Fish shape
صيداوي	Sedawi	From Saida (town in Southern Lebanon)
صفري (تموزي)	Sfaery (Tamozi)	Summer (July)

English translations provided by Ali Rida Issa.

Fig genetic resources

Most fig cultivars are derived from old local selections and maintained by vegetative propagation. Hundreds of variety names are listed in the literature. The various traits used in germplasm descriptors (IPGRI and CHIEAM 2003) include fruit, agronomic and technological characteristics, as well as molecular markers that are used for variety differentiation and germplasm description. Morphometric and molecular analysis revealed high genetic diversity within cultivated fig germplasm (Mars 2003). Numerous collections have been established in several countries (WIEWS 2002). The European Central Minor Fruit Trees Database is maintained by the Horticulture Department, University of Florence, Italy, and it holds information on 432 accessions of fig, among other underutilized fruit tree species, collected both *in situ* and *ex situ* by European institutions of France, Greece, Italy and Spain. The Directorate of Scientific Agricultural Research has been collecting varieties to be included in a field gene bank (ICARDA 2001). A guidebook to the local cultivars of figs of Syria is in preparation.

The System-wide Information Network for Genetic Resources (SINGER), an information exchange network of the centres supported by the CGIAR and associated partners, lists a number of collections of *Ficus carica* germplasm (See Table 6)

JUJUBE

Jujube (*Ziziphus jujuba* Mill.) is a wild fruit tree or shrub, also called Chinese jujube, and it is cultivated in drier parts of China and to a limited extent in Armenia, Azerbaijan, France, Kyrgyzstan, Spain, Syria, Turkmenistan, south-western USA, Uzbekistan and Yugoslavia (ICUC 2001). The jujube is a particular good tree for dry regions because it can withstand long periods of drought and is adaptable, withstanding extremes of both hot and cold temperatures (ICUC 2002). The tree can also tolerate many different soil types: high salinity, high alkalinity, arid and waterlogged. Its yellow flowers bloom in summer and the small, round, dark-red fruits of the jujube tree are harvested in October in the Northern Hemisphere. The fruits do not ripen simultaneously and can be picked over the course of several weeks. Jujube trees fruit without cross-pollination. Cultivated trees need moderate watering during the growing season to ensure an optimum fruit yield. Neither fertilization nor any other special care is necessary. Unpruned trees produce as much fruit as those that have been pruned. Extensive winter pruning, however, helps to keep the plants in better health and allows for easier harvesting by making the fruit more easily reachable.

Table 6. *Collections of* Ficus carica *germplasm reported in the SINGER system.*

Institute	No. of Accessions	City	Country
Research Institute of Fruit Trees and Vineyard	12	Tirana	Albania
Station Experimentale, Inst. Techn. de Arboricult. Fruit. et de la Vigne	58	Boufarik	Algeria
Asociación ANAI	1	Sabanilla, Montes de Oca	Costa Rica
National (CYPARI) Genebank, Agricultural Research Institute	39	Nicosia	Cyprus
CIRAD-FLHOR Station de la Guadeloupe	6	Capesterre Belle-Eau	France
Conservatoire Botanique National de Porquerolles	143	Hyeres	France
Greenhouse for Tropical Crops, University of Kassel	no data	Witzenhausen	Germany
Olive, Fruit and Vegetables Institute of Kalamata	52	Kalamata	Greece
Caribbean Agricultural Research & Development Institute (CARDI)	1	St. George's	Grenada
Centro Universitario Regional del Litoral Atlantico (CURLA), UNAH	1	La Ceiba, Dept. de Atlantida	Honduras
Indian Institute of Horticultural Research	23	Bangalore, Karnataka	India
Istituto Sperimentale per la Frutticoltura	250	Rome	Italy
CNR - IVALSA	5	Sesto Fiorentino (FI)	Italy
Agricultural Research Centre (ARC)	8	Tripoli	Libya
Estacao Nacional de Fruticultura Vieira Natividade	50	Alcobaca	Portugal
Direccao Regional de Agricultura do Algarve	60	Faro	Portugal
N.I. Vavilov All-Russian Scientific Research Institute of Plant Industry	180	St. Petersburg	Russian Federation
Junta de Extremadura. Servicio de Inv. y Desarrollo Tecnológico Finca la Orden	202	Guadajira	Spain
Instituto Canario de Investigaciones Agrarias (ICIA)	20	La Laguna, Tenerife	Spain
Conselleria d'Agricultura i Pesca	52	Palma de Mallorca	Spain
Centre de Mas Bove, Inst. Recerca i Tecnologia Agroalimen. (IRTA)	92	Reus, Tarragona	Spain
Almonds Research Unit, Department of Horticulture, ACSAD	108	Damascus	Syrian Arab Republic
Arab Centre for the Study of Arid Zones and Dry Lands (ACSAD)	370	Damascus	Syrian Arab Republic
Chiayi Agricultural Experiment Station, TARI	1	Chia-yi	Taiwan, Province of China
Department of Horticulture, Kasetsart University	18	Bangkok	Thailand
Fig Research Institute	no data	Aydin	Turkey
National Germplasm Repository, USDA, ARS, University of California	141	Davis, California	United States of America

Jujube wild trees, Lattakia Province, Syria.

In Syria, jujube grows mainly in the coastal mountain areas of the Mosyaf and Akrad mountains, situated in the governorates of Lattakia and Tartus. Some wild trees reach 200 years of age. The tree is deciduous, losing its leaves during the colder months. Cultivation of the jujube tree is generally limited to home gardens in the Lattakia and Tartus areas, with some experimental growing elsewhere. This tree is not usually grown for commercial production. Before 1967, cultivated trees could only be found in two villages in Syria: Al-Rabie and Fadre (Lattakia). Its cultivation then spread to other regions, such as Homs and Tartus, and to some areas in Lebanon. Nurseries have been sending 2 000 saplings to these regions each year. The most important area for jujube cultivation has been that of Al-Rabie, situated in Al Haffeh (near Lattakia), with about 4 000 trees planted in the area by some ten farmers (ICARDA 2000). These trees have yielded more than 100 t of fruit annually. According to farmers, a single jujube tree had the same economic value as that of 10 to 20 apple trees. The farmers also stated that the jujube tree was more valuable than any other fruit tree growing in the region, since it did not require any special care. In other areas, farmers cultivated trees in their home gardens for both their own use and for additional earnings through sale at harvesting time. Jujube trees begin to fruit 3 years after planting, and 7 years after planting they can produce as much as 100 kg of fruit a year, with an average annual yield of 50 kg.

Jujube fruits, Syria.

Local names and uses

Jujube is commonly called *annab* (عناب) everywhere in Syria. Growers and collectors recognized only a single variety of cultivated or wild-harvested jujube tree. Fruit—both fresh and dried—are eaten. They are rich in vitamins C and A, and the B-complex. Ripe fruits taste like green apples, while over-ripe or dried fruits have a taste similar to dates. Fruit from the jujube tree is eaten only in the coastal areas of Lattakia and Tartus.

In traditional medicine, the leaves are used to alleviate liver troubles, asthma and fever. Fresh fruits are applied to cuts and ulcers. In addition, they are mixed with salt and chilli peppers to aid indigestion and similar ailments. Dried fruits are rich in vitamin C and are used as remedies for throat infections, pulmonary ailments, flues and fevers. Dried ripe fruit is a mild laxative. Seeds have a sedative effect and are taken, sometimes with buttermilk, to alleviate nausea, vomiting and abdominal pains during pregnancy. Root bark juice is believed to alleviate the symptoms of gout and rheumatism. As animal food, the leaves are readily eaten by camels, cattle and goats, and are considered nutritious. They are also gathered as food for silkworms. As concerns other uses, the flowers are a minor source of nectar for honeybees. The honey is light and has a delicate flavour. Trees are also cultivated in home gardens as ornamental plants.

Jujube genetic resources

In South and South-East Asia, there is rich diversity in jujube (Arora and Ramanatha Rao 1998); and there are ten well-described varieties of *Ziziphus jujuba* in South Asia (ICUC 2001). During the 1960s, India used this diversity to produce varieties suitable for commercial cultivation. There are about 700 lines of *Z. jujuba* in China, currently under evaluation for improvement (Mengjun 2003). In Syria, jujube germplasm is locally obtained. There are two national nurseries, located in Lattakia and Tartous Provinces, that provide 2000 local saplings to these regions each year. This germplasm has not been collected for characterization, and there is little research currently underway in Syria.

The System-wide Information Network for Genetic Resources (SINGER), an information exchange network of the centres supported by the CGIAR and associated partners, lists a number of collections of *Ziziphus jujuba* germplasm (See Table 7).

Table 7. *Collections of Ziziphus jujuba germplasm reported in the SINGER system.*

Institute	No. of Accessions	City	Country
Asociación ANAI	1	Sabanilla, Montes de Oca	Costa Rica
Regional Station Shimla, NBPGR India	2	Simla, Himachal Pradesh	India
National Genebank of Kenya, Crop Plant Genetic Resources Centre, KARI	1	Muguga	Kenya
Institute of Plant Breeding, College of Agriculture UPLB	4	College, Laguna	Philippines
Chiayi Agricultural Experiment Station, TARI	7	Chia-yi	Taiwan, Province of China
Fruits Crops Research Centre	4	Vinh phu	Viet Nam

LAUREL

Laurel (*Laurus nobilis* L.) is an evergreen forest tree of the Laureaceae family, which grows wild along the coastal area of Syria, to which environment it is ideally suited. It grows wild, generally above 200 m. Laurel is known for its resilience to pests and abiotic stresses. It is also resistant to extreme temperatures and tolerant of both coastal conditions and warm climates. Laurel trees flower around mid-April. Its fruits are small and round (or oblong), with very dark berries that ripen between October and December. In 2002, there were 1 097 laurel trees covering about 1 374 ha (about 3% of the forestry fruit area) in Syria (MAAR 2002). These trees can be found, in particular, in the mountain areas of Kessab (Lattakia Govenorate) and Kadmus (Tartus Govenorate). In Syria, Laurel is an example of the wild biodiversity of a semi-arid environment that has been exploited by rural and forestry communities for centuries, using their indigenous knowledge. Well known, traditional products—laurel leaves, laurel oil (and laurel butter) and laurel soap—have been produced in response to great demand by the local market, demand that has remained stable for decades. Now, export demand for these products has been growing and there are increasing opportunities for realizing the value of these products through the market.

Laurel landscape, Lattakia Province, Syria.

Local names and uses

Laurel is commonly called *ghar* (غار) everywhere in Syria. Growers and collectors formally recognize only a single variety of cultivated or wild-harvested laurel. However, collectors say that the quality of laurel oil differs according to its fatty acid content (particularly C_{10} and C_{12}), which varies according to the genetic character of the laurel used. Each variety is in fact characterized by a well defined type of berry, differing in scent, size and colour.

Box 1. Ghar soap history

Laurel oil, and the soap made with it, have a very old tradition in Syria, mentioned in legends. Tradition speaks of the 'magic' properties of laurel oil. Legend says that the Egyptian Queen Cleopatra and the Syrian Queen Zenobia kept their skin and hair fresh and shining by using ointments made out of laurel.

Although olive plantations and olive oil production have been known since antiquity (tablets from Ebla, dated 2400 BC, mention olives and olive oil), the actual manufacture of soap from oil seems to have been a more recent development.

Soap is first mentioned as a medicinal lotion used in the treatment of certain conditions of the scalp and skin. This early soap was produced from animal and vegetable fats and oils. It is cited in early Sumerian and Assyrian tablets, as well as in Egyptian papyri. The Roman historian Pliny, writing in the 1st century AD, attributes the invention of soap to the Gauls.

Whatever its origins, by the Early Middle Ages soap produced from laurel oil had become a thriving business all over the civilized world. Aleppo in northern Syria was particularly famous for the quality of its soap, produced in the many small workshops concentrated in the Bab Qinnisrin area.

By the 16th century, the workshops had been transformed into large factories and many specialized areas (Hay Al-Masabin), souks (Shari Al-Sabbana) and khans (Khan Al-Saboun) came into being to cater for the booming soap industry. And they remain in existence today (SEBC 2003).

Leaves

Laurel (also called bay) leaves are used extensively in Syria, most Mediterranean countries and wider, for cooking, used either fresh or after the 'drying' process. They are used to flavour soups, stews, pickling brines, sauces, marinades and poultry. The leaves, in addition, are considered to have medicinal properties. Dried leaves are brewed as a herbal tea and are used in herbal medicine to treat rheumatism, joint pains, schizophrenia, stress, to stimulate the appetite and as a sedative. It is also possible to extract an aromatic essential oil from the leaves, which is used for medicinal purposes. This process, however, has not yet been adopted in Syria.

Oil (extracted from laurel tree fruit)

Laurel oil and laurel butter are used in cosmetics as a body balsam for moisturising and massaging. It is also used as a hair strengthener. In herbal medicine, the oil is used to treat itching skin, earache, asthma and urinary ailments. A cream made with laurel oil and beeswax is used to treat skin irritations and wounds.

Laurel berries, Syria.

Soap

Laurel oil is mainly used in the production of laurel soap. Laurel soap is known for its unique perfume and is used for both body and hair cleansing. It has distinctive qualities, such as skin nourishing, softening, refreshing, cleansing, protecting, deodorizing and antisepsis. It is also used for sensitive and damaged skins.

Box 2. Homemade laurel soap recipe

Ingredients:

6 kg olive oil

1 kg laurel oil

1.5 kg soda

Dissolve the soda into the water at a ratio of 1 soda : 2 water (v:v). Mix the olive and laurel oils, warm on fire. Add soda solution gradually and stir until the mixture turns dough-like through the evaporation of water. Shape in moulds and leave to dry in the sun.

Laurel genetic resources

Some ecogeographic studies on the genetic diversity of *Laurus nobilis* have been carried out in Spain, France and Italy. Samples have been analyzed using molecular characterization (Arroyo-García 2001). *Laurus nobilis* germplasm is maintained by three institutes, according to information available, namely Banco Português de Germoplasma Vegetal, Portugal (WIEWS, 2002), Depart. de Proteccao de Plantas Univ. Tras-os-Montes e Alto Douro, Vila Real Codex, Portugal (SINGER database), and by the Greenhouse for Tropical Crops, University of Kassel, Witzenhausen, Germany (SINGER database). In Syria, laurel germplasm has been neither collected nor characterized. The germplasm available is from local wild plants and one government nursery. Initial studies on characterization of the laurel varieties are envisaged at the University of Damascus.

CAPER

Caper (*Capparis spinosa* L.) is a spiny perennial shrub. It grows up to 1 m in height, has trailing branches and orbicular leaves, with stipules modified into spines. White flowers blossom in summer. The caper is adapted to dry heat and intense sunlight and can survive in temperatures of over 40°C (Alkire 1998). Caper cultivation can be found in most Mediterranean countries, where the plant grows wild as well. Capers probably originated from dry regions in west or central Asia. Known and used for millennia, capers were mentioned by Dioscorides as being a marketable product of the ancient Greeks (Alkire 1998). Plants are grown from seed and by vegetative cuttings. A full yield can be expected in 3 to 4 years. Plants are pruned back in winter to remove dead wood and sprouts. Pruning is crucial to high production. Heavy branch pruning is necessary, as flower buds sprout on the young, first-year branches. Three-year old plants yield 1–3 kg of caper flower buds per plant. Varieties are selected for spinelessness, round firm buds, and flavour. In Syria, caper can be found growing wild everywhere, around dry and rocky areas, by roadsides and on old walls. The wild plants commonly referred to as *C. spinosa* belong either to *C. sicula* or *C. orientalis*. Other species can be found in Syria, such as *C. aegyptia* in the Damascus basin and in the Upper Syrian Euphrates region (Riviera et al. 2003). Capers are found growing wild in particular abundance in four main areas in Syria: (1) Eufrate, east Aleppo; (2) northern Aleppo; (3) Alsalamie area (Hama); and (4) Al Jazira (north-east). Caper seeds have been recovered from archaeological sites in northern Syria in Mesolithic layers (Hillman 1975). In Syria, caper occurs only as a wild species and is cultivated only on an experimental basis in research nurseries.

Caper plant in Salamie, Tartous Province, Syria.

Caper flower, Syria.

The main caper product for trade in Syria is the young flower bud, collected before the flowers have formed. After collection these buds are preserved in salt, or pickled in vinegar. The capers are used in Europe and other countries as a food condiment. The fragrance is spicy and a little sour, while the taste is slightly tart. Caper fruits ('cucumber') are used for commercial purposes in other countries. There is little use and market for the plant in herbal medicine in Syria. although caper products are sold in herbal shops in Syrian towns and their use is known by herbalists. The flower buds and fruits are not used as a food in Syria. However, a jar containing carbonized flower buds and unripe fruits at Tell es Sweyhat in Syria, may indicate the established use as food of caper pickles in the Bronze Age (Rivera et al. 2003).

Local names and uses

Caper is commonly called *kabar* (قبار) everywhere in Syria.

'Caper' and its relatives in several European tongues can be traced back to the Greek *kápparis*, whose origin (as that of the plant) is probably West or Central Asia. It is interesting to note that, in Syria, the Arabic word *alkabar* (the caper) refers to everything that surrounds the *alkebr* (=tomb or grave) (Issa, pers. comm.). The botanical specific epithet *spinosa* (=thorny) refers to the many sharp thorns of the wild plant, which are absent in some cultivars. Collectors and traders in Syria formally recognize only a single variety of wild-harvested caper plants.

Neither the fruits (cucumbers) nor the flower buds (capers) are commonly eaten in Syria. Very rarely, the capers are used as pizza toppings, or for adding flavour to smoked salmon sandwiches in a few restaurants in Aleppo and Damascus. The caper plant has had various uses in herbal medicine in Syria, although less used in recent times. Dried leaves in vinegar are applied to ulcers and scabs on the head. Root bark powder is used to cure rheumatism (when mixed with olive and other oils) and diseases of the urinary tract. It also has diuretic properties. Flowers (dried and boiled in water) are used to alleviate the symptoms of asthma. Flower buds are eaten to strengthen the body and as an aphrodisiac. The nectar and the sugars contained in the heart of the flower are taken by bees that produce caper honey. This is an expensive and precious honey blend, selling for LS 1000/kg. Due to its bitter taste, it is not used for eating, but it is used to treat rheumatism and diseases of the urinary tract (e.g. kidney stones). Caper bushes are used as animal feed: camels, goats and sheep graze the plant.

Caper genetic resources

According to Lawrence (1951), the genus *Capparis* includes 350 species; The wild plants commonly referred to as *C. spinosa* belong either to *C. sicula* or *C. orientalis*. Other species can be found in Syria, such as *C. aegyptia* in the

Damascus basin and in the Upper Syrian Euphrates region (Riviera et al. 2003). There is a lack of information on the genetic diversity of caper germplasm (Khouildi et al. 2000). Most capers are gathered from natural habitats with little attention to the preservation of germplasm resources and genetic diversity. There is a need to carry out preliminary studies to assess the extent of genetic variation between and within natural populations of caper, as well as to assess techniques for sustainable harvesting and further adaptation of domesticated varieties. Some investigation and characterization of caper genetic diversity has been done, including molecular characterization of populations of Italian and Tunisian capers (Khouildi et al. 2000). Studies on the biotypes commonly cultivated on the Italian islands of Salina and Pantelleria were carried out by Barbera et al. 1991).

The following institutions have specific caper germplasm collections:
* Instituto di Coltivazioni Arboree, University of Palermo, Italy.
* Instituto di Coltivazioni Arboree, Universita' di Napoli Federico II, Naples, Italy.
* National Research Council (CNR) Germplasm Institute, Istituto di Agronomia Generale e Coltivazioni Erbacee, Bari, Italy.
* Instituto de Agricultura Sostenible, Consejo Superior de Investigaciones Cientificas, Cordoba, Spain.

The SINGER System reports that *Capparis spinosa* germplasm is also maintained in two other gene banks: Greenhouse for Tropical Crops, University of Kassel, Witzenhausen, Germany, and Suceava Genebank, Romania

Commercial seeds and plant sources are available in Australia, Italy, UK and USA (Alkire 1998). In Syria, there has been neither collection nor characterization of caper germplasm. The germplasm used in Syria derives from local wild plants. Small studies to characterize the genetic diversity of this species in Syria are just beginning at the University of Aleppo.

PURSLANE

Purslane (*Portulaca oleracea* L., Family Portulacaceae) is an annual, herbaceous plant, with a branched, fairly ascending stem of up to 50 cm, and with leaves that are fleshy and smooth. The flowers are yellow. The fruit is in a capsule. Purslane is drought tolerant.

Purslane grows wild in almost any sunny area, including flowerbeds, cornfields and waste places, and it is suitable for both cold climates as well as warm areas. It is a source of Vitamins A, C and E, as well as containing an omega-3 fatty acid (Simopoulos 1992). In Syria, as in many countries around the world, it is also cultivated. Seeds maintain their germinating capacity for 8–10 years, and their viability is extended if they are stored dry at a low temperature. Purslane is collected from wild areas in many parts of Syria, or from the cultivated fields where it grows as a weed. It is mainly cultivated, either outdoor or in greenhouses. Greenhouse cultivation is mainly in Damascus.

Wild purslane growing around a potato field in Damascus Province, Syria.

Local names and uses

Purslane is commonly called *bakla* (بقلة), *hamga* or *riyla* in different areas in Syria.

Collectors in Syria recognize only a single variety of wild-harvested purslane plants. Growers and traders distinguish two sorts according to their harvesting time: *rabieh* for the spring varieties, and *hamga* for the summer one, but according to studies from the University of Aleppo, the genetic diversity is very low. Consumers, however, distinguish cultivated from wild purslane.

The fresh leaves are eaten raw in salads, in particular in the traditional oriental cuisine salad called *fattoush*. Leaves are used to flavour yoghurt, together with mint and garlic. There are other traditional recipes, where leaves are cooked with green onions and lamb, or with eggs when families cannot afford to buy meat. The leaves are cooked in soups, or cooked in olive oil and eaten with onion and bread. As a medicinal plant, it is considered to have anti-scorbutic, diuretic and cooling properties. Being rich in mineral salts and with a high water content (95%) and mucilage content, it has emollient and smoothing properties for irritations of the bladder and urinary tract. The juice was supposed to stop haemorrhages. The seeds, bruised and boiled in wine, were given to children as a vermifuge. Seeds are used traditionally to treat the 'evil eye'.

Box 3. Fattoush salad recipe

Fattoush is a typical dish eaten as a starter in Middle Eastern cuisine, and commonly eaten in all parts of Syria, either at home or in restaurants, and known in other countries. It is very common, in particular during Ramadan festivities

Ingredients:

2 – 3 tomatoes, cubed

200 g purslane leaves

3 small cucumbers, chopped

1 diced medium green pepper

1 chopped onion

½ small lettuce, shredded

chopped parsley

chopped fresh mint

1 flat bread, fried and cut into pieces

Dressing made from equal amounts of olive oil and pomegranate juice, and seasoned with salt

---

Purslane genetic resources

Globally, *Portulaca oleracea* germplasm is maintained by the following institutes (WIEWS 2002):
* Conservatoire Botanique de Ressources Génétiques de Wallonie, Belgium.
* Safeguard for Agricultural Varieties in Europe, Switzerland.
* Millennium Seed Bank Project, Seed Conservation Department, Royal Botanic Gardens, Kew, UK.
* Research Institute of Forest and Rangelands, Iran.
* National Genebank of Kenya, Crop Plant Genetic Resources Centre, Kenya Agricultural Research Institute.
* Departamento de Botanica e Engenharia Biologica, Instituto Superior de Agronomia, Lisbon, Portugal.
* Medicinal and Aromatic Plants Research Station Fundulea, Romania.
* Centro Nacional de Conservación de Recursos Fitogenéticos, Ministerio de Agricultura, Venezuela.
  In addition to the above collections, the SINGER system reports:
* Banco Base Nacional de Germoplasma, Instituto de Recursos Biológicos,INTA, Castelar, Prov. de Buenos Aires, Argentina.

- Centre for Genetic Resources, The Netherlands (CGN), Wageningen, The Netherlands.
- Horticultural Research Section, Agricultural Research Corporation, Wad Medani, the Sudan.
- Plant Genetic Resources Department, Aegean Agricultural Research Institute, Izmir, Turkey.

In most cases there are only 1 or 2 accessions in each gene bank. The genetic diversity has yet to be well studied. In Syria, purslane germplasm is not maintained in any collection, and the germplasm available derives from local wild and cultivated plants. There is currently no research characterizing the genetic diversity of this species in Syria.

MALLOW

Mallow (*Malva silvestris* L., Family Malvaceae) is widespread in Europe and in parts of West Asia, Syria, Iraq, Jordan and North Africa, especially Morocco. It is most abundant in warm countries. This plant is characterized as a herbal annual or perennial low shrub with ovoid leaves that are thick, succulent and salty. The flowers are pale lilac, small and hermaphrodite.

Mallow is distributed everywhere in Syria on waste ground, field verges and roadsides. Until a decade ago, mallow was found only wild in Syria, but now it is also cultivated. The leaves of mallow are traditionally eaten in Arab countries. Mallow is used also for medicinal purposes, as well as a dye, fibre source and for litmus.

Mallow flower, Syria.

Local names and uses

Mallow is commonly called *khebbezeh* (خبيزة) everywhere in Syria.

The cultivated variety has bigger leaves than the wild plants. The leaves are eaten raw or cooked. The young raw leaves make a rare but acceptable substitute for lettuce in a salad. The fresh leaves and the branches (when they are small), alone or mixed with *Brassica arvensis, Taraxacum officinale* and other herbs are eaten cooked, as a traditional dish in country areas. This dish, called *khebbese* or *marshushe,* is being forgotten among families living in towns. This dish is only eaten during the collection period, in March–April. The flowers are not used in Syria, while in other countries they are used as a garnish in salads.

All parts of the plant are used in herbal medicine, since mallow has astringent, demulcent, diuretic, emollient, expectorant and laxative properties. A tea is usually made from dried leaves and flowers to treat respiratory system diseases and problems with the digestive tract. When combined with eucalyptus, it makes a good remedy for coughs and other chest ailments. The leaves and flowers in a poultice are used for bruises, swelling, inflammations, insect bites and stings, etc. The plant is an excellent laxative for young children. Local people know the medicinal properties of the plant and they use it as a natural healer.

Box 4. Recipe for mallow stew

This recipe for Marshushe—mallow stew—is from the Jabal al-Hoss area, northern Aleppo Governorate.

Ingredients:

Leaves and tender braches of mallow (wild or cultivated)

2 Onions

4 spoons of burgel (bulgur or burghul)

4 spoons of olive oil

Cut the leaves and the branches and cook them in olive oil with the onions. Add the burgel and stir. Then cover and leave the soup until the burgel is cooked.

Mallow genetic resources

Globally, *Malva sylvestris* germplasm is maintained by the following institutes (WIEWS 2002):
- Genebank Department, Vegetable Section, Olomouc, RICP Prague, Czech Republic.

- Millennium Seed Bank Project, Seed Conservation Department, Royal Botanic Gardens, Kew, UK.
- Faculty of Agricultural Sciences, Pannon University of Agriculture, Hungary.
- Plant Breeding and Acclimatization Institute and Institute of Medicinal Plants, Poland.
- Departamento de Protecção de Plantas, Universidade Tras-os-Montes e Alto Douro, Portugal.
- Medicinal and Aromatic Plants Research Station, Fundulea, Romania.
- N.I. Vavilov All-Russian Scientific Research Institute of Plant Industry, Russian Federation.

In addition to the above, SINGER reports a *Malva sylvestris* collection at Federal Centre for Breeding Research on Cultivated Plants (BAZ), Braunschweig, Germany.

In Syria, mallow germplasm has neither been collected nor characterized. The germplasm available is from local wild and cultivated plants. There is no current research in Syria looking at the genetic diversity of the species in the country.

5

Findings: species and market chains

The results reported in this chapter come from focus group findings and their analysis. Checklists were used in focus group interviews with academics, researchers, farmers and traders to obtain information about the actors involved in the chain, their links, and the structure of the market chain. The results are reported for each of the six species studied.

Following the value-chain approach, a map was plotted of the chain actors and flow of the product from collectors and growers to retailers, as well as the distribution of added values along the chain. Issues related to the quality of the product were reported. Some aspects related to the market environment are described, including the organization of the market, the impact of policies, and legal issues.

Finally, the constraints identified and opportunities for improvement in the marketing of the products derived from these species are listed. The constraints and opportunities are summarized by species in a matrix to show similarities and contrasts among the species.

With regard to the distribution of species, figs are grown, processed and traded in all the surveyed sites. Jujube is only grown and collected in the central and western (coastal) regions. Capers are only found wild, and processed and traded in northern, central, eastern and southern sites. Laurel trees are grown and found wild in the western site; berries and leaves are processed in the northern and western sites and traded everywhere. Mallow is collected at all the sites, but grown only in eastern and southern sites; it is not processed and it is traded at all sites. Purslane is grown, collected and traded unprocessed in all areas.

FIG

Mapping the value chain

The value-chain mapping for Kafranbel area, Idleb Governorate, is shown in Figure 3. Farmers came to the main wholesale market (*souk-al-hal*) to sell their fresh figs to wholesalers. The farmer generally dealt with the same wholesaler. This kind of commercial relation was quite stable and implied a good level of trust between the two actors. At the main wholesale market, middlemen bought the products from the wholesalers, and tried to make some money re-selling them to retailers or to consumers on the main wholesale market or on the sub-main market. The price depended on demand and on the variety, and varied between LS 20 and 35/kg. Wholesalers sold the product on the market on behalf of the farmer, retaining a 5% commission. Middlemen sold the product adding 3–5% to the price. Finally, the retailers' price to final consumers was about LS 25–55/kg. When supply much exceeded demand, the price could decrease to half the normal price.

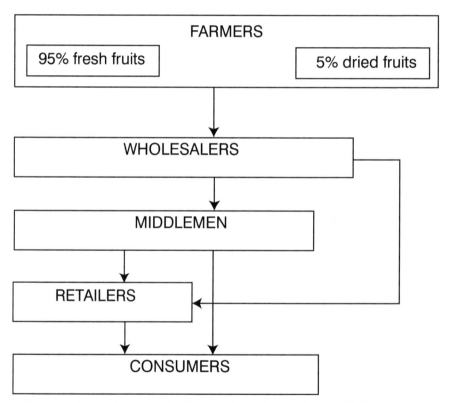

Figure 3. *Value chain for the fig, based on observations in Kafranbel area, Idleb Governorate, Syria.*

Processing of dried figs

The figs start a natural drying process on the trees. After harvesting, farmers lay the figs on rocks, elevated from the soil, to complete the drying process.

   In the past, farmers stored the dried figs under rocks, helping prevent the entry of insects into the dried figs. This method required some farmers to stay awake during the night to look after these natural outdoor stores. Nowadays, dried figs are stored in storerooms at the farms. Other fig varieties, for example *mistah*, are picked from the trees when they are mature, then the fruit is opened and allowed to dry in the sun. They are then transformed into fig paste by assembling them, softening them with steam and then pressing them again. This drying process minimizes insect infestation during the storage period (Al Ibrahim 1997).

Fig varieties in the Aleppo souk, Syria.

Quality

There are about 56–60 figs of the *baiadi* variety to a kilogram. This corresponds to *extra class,* (size no. 5) according to the classification used for exportation to Europe. The *baiadi* variety is suitable for drying. The natural process used to dry the fruit and storage in storerooms allows insect access to the dried fruits and reduces the quality of the product, so that product does not satisfy the quality standards required for export.

There are about 90 figs/kg in the *mistah* variety. This corresponds to Class 1 (size no. 9) according to the European classification. Such figs are small and unappealing to consumers for either fresh or dry consumption, and they are processed into fig paste, which can be conserved free from insects without the use of preservatives. Fig paste is suitable only for the local market.

Market environment

The market for fig was a free market in Syria and there were no price regulations. The price was established by the supply:demand balance. During the 1970s and 1980s, there was a reduction in production, reflecting a general decrease in the fig price. The farmers decided to instead increase production of the more profitable olive oil. From the early 1990s, the price of olive oil started to fall because of greater supplies from both domestic and international markets, while the price of figs increased again, and so did production. Recently, the price of olive oil had been about LS 160/kg (US$ 3), while the price of fig was LS 10/kg (US$ 0.20).

In the filière approach, the fig production system can be considered a 'market' mode, as the exchange was mainly done within a marketing channel, through commercial intermediaries, on the basis of a competitive price. The dominant actors were market actors, such as small-scale producers, intermediaries and private operators.

Constraints and opportunities

Storage problems

There was no use of chemicals, either during the production cycle or for processing of figs. The drying process was a natural process—figs were naturally dried by the sun—and this inefficient natural process created problems during storage of the dried figs, namely:
- development of fungus, due to the high residual moisture; and
- insect infestation in the dried fruits.

Farmers used a traditional method to kill insect eggs in the dried figs, which was to pour boiling water onto the already-dried figs, and then to dry the figs again. Because of this process, the figs had a darker appearance and a higher moisture level and could not be sent to the traders, as they had lost the characteristics required by final consumers, namely a pale colour and low moisture content (<22%). Hence, this method was applied only to the figs for home consumption. The farmers lacked knowledge of processing technique that would provide good conservation of the product, while traders were aware of the technique.

Lack of market transparency

Farmers had access to the market through only a few traders. These traders fixed the price to their own advantage (i.e. very low compared with the selling price), since they knew that farmers had no opportunity to sell their figs to other traders or markets. In addition, traders used a better process to prevent insect infestation in the dried figs and to facilitate product conservation. Farmers did not know the final consumers and so were obliged to sell their product to the traders, since they had no alternatives (they simply lost the figs, as they could be stored for only a very short time).

Lack of support

There were neither governmental subsidies nor help from international organizations to support the development of the production and processing of figs in this area. There was a lack of technical support in-country.

Opportunities

The product grown in this area has good potential in the development of a greater market or to reach a niche market. It could be considered a 'green' product, as there has been no use of chemicals in the production and processing. The quality of the product can be very good, as well as the average annual production. Not least, continued production contributes to conservation of the species biodiversity.

Some actions proposed by the representatives of the Extension Unit for Agriculture in Kafranbel, Directorate of Agriculture of Idlib, to increase market value were to:

* provide farmers with technical support for cultivation;
* build up associations of farmers in order to be able to influence market price; and
* ask the Ministry of Agriculture about the possibility of organizing a training experience in Turkey for farmers from Kafranbel, where they could learn alternative processing practices to better conserve the product.

JUJUBE

Mapping the value chain

Farmers living in Lattakia and Tartus both collected fruit from wild jujube trees and grew a few trees in their home gardens. Almost all (95%) of the fruit from both wild and cultivated trees was sold as fresh fruit. Ripe fruits could be stored at room temperature for about a week. Fresh jujube berries were not available in local markets, but were offered for sale by subsistence farmers, or collectors, along the highways, along the coast and sometimes across the Lebanese border. Traders bought fresh jujube fruit from the farmers and collectors, and then sold the fruit direct to consumers on the highways in the coastal areas or exported the product to Lebanon, where they sold it to wholesalers. A few traders sold fresh jujube fruit to restaurants in the area. The price was determined by the market and it was higher at the beginning of the harvesting season, in September. The price to the final consumers of fresh fruit varied between LS 100 and LS 300/kg. The price of fresh jujube fruit also changed according to the difficulty in exporting to Lebanon, where taxes were applied to products originating in Syria. Farmers received LS 50–150/kg from traders that came and bought at the farms. The traders' commission was quite high, as they managed to sell the produce to consumers at LS 100–300/kg. Plans to process fresh jujube fruit into jam had been initiated by farmers in Al-Rabie (ICARDA 2000), and contact had already been made with jam factories in Damascus.

The remaining 5% of the harvest was dried by farmers. Usually, the jujube fruits that were dried were of lesser quality or from late in the harvest season. Fruits were air-dried outside, variously in direct or indirect sunlight, and then stored on the farm. Fruits could also be allowed to dry on the tree and be left there for a limited period. Each 5 kg of fresh fruit gave 1 kg of dried fruit. Dried fruit was used for household consumption or was used as gifts to families in the community.

Jujube for eating in Tartous Province, Syria.

It was also sold to traders, who then sold the dried fruit to herbal shops in major towns, such as Aleppo and Damascus. The dried fruit was sold for LS 700–1000/kg. Figure 4 shows the jujube value chain.

Quality

There was no quality control for the fruits. There was no formal training available for cultivation or processing of jujube in Syria.

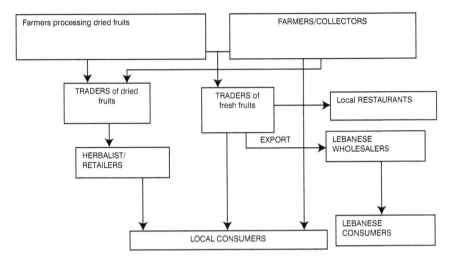

Figure 4. *Value-chain for jujube in Syria.*

Market environment

The market for fresh and dried jujube fruit was an open market and the price was determined by the demand and supply balance for the product. At the very beginning of the harvesting season (early September), the price was higher. The price fluctuated according to the Lebanese tax regime to which the fruit was subject, but export to Lebanon was frequently by unofficial routes.

From the point of view of the filière approach, jujube production could be considered as a 'domestic' mode, as the actors involved in the chain belong to family lineages and fragile groups. Family members from rural communities were the ones in charge of the production, processing (very rare) and commercialization, using very traditional techniques despite suitable industrial ones being available. The exchanges are made at a very informal level, including household-level exchanges.

Constraints and opportunities

Constraints

- There was very limited local consumption, due to lack of awareness about the jujube and its fruit.
- The fluctuation of the price of fresh fruit in Lebanon, where jujube fruit was mainly sold, was due to the erratic application of Lebanese import duty regulations.

- There was a lack of proper packaging for the fruit, leading to damage during transport before sale.
- Quality control for the fruit was absent.
- There was a lack of research on the jujube tree and its cultivation, and a lack of formal training on the cultivation of jujube trees and the processing of its fruit.

Opportunities

- Jujube fruit jam represents a potential new product for the Syrian market, with some export potential. An Agrobiodiversity Food Products Fair was held in 2000 within the framework of the project on Conservation and Sustainable Use of Dry Land Agro-biodiversity, funded by GEF/UNDP. Samples of jujube jam were tasted at the fair. Samples were also distributed to hotels in Damascus, Tartus and Lattakia, in 2002, in order to test the market (with good results).
- The taste of jujube jam is similar to that of date, which is popular in Syria and neighbouring countries, and so should easily appeal to consumer tastes in those countries. Jujube jam is extensively used in the pastry industry in other countries, such as China.
- Dried fruit production has the advantage that the fruit can be dried *in situ* on the tree.

LAUREL

Mapping the value chain

Leaves and berries were collected from wild laurel trees by members of forestry communities, with collection controlled under Law No. 7 of 1994, through regulations enacted by the Ministry of Agriculture and Agrarian Reform (MAAR), which identified the laurel tree as a *herage* (forest) species. A few farmers grew laurel trees in their home gardens. The value chain involved: collection and trade of laurel leaves; collection of laurel berries, their processing into oil and its trade; and laurel soap processing and trade. These are summarized in Figure 5.

Leaves were dried naturally by the collectors, who then sold them in the markets or to traders for LS 80–100/kg. The traders sold the laurel leaves to herbal shop retailers or to foreign traders. The price for local consumers was about LS 100–300/kg. Laurel leaves were sold in the herbal market without packaging. Very few collectors harvested both leaves and berries.

Laurel berries were collected by forestry community members of Kessab and Kadmus, who also extracted the oil. The fruit contains 2–3% of essential oil. The pulp yields 25% laurel oil or 'Laurel butter' (Al-Hakim 1994). The extraction was a manual process. First, the whole berries were boiled in water for 6–8 hours in a metal container heated on a fire, with stirring. The oil rose to the surface and was skimmed off with a wooden spoon. This liquid was then filtered and bottled.

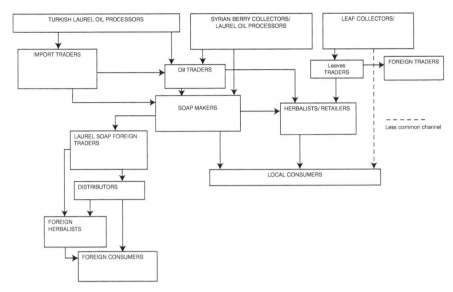

Figure 5. *Value chain of laurel-derived products in Syria.*

Laurel leaves consumer in the Damascus souk, Syria.

The yield from 16 kg of laurel berries boiled in 60 litres of water was about 10 litres of laurel oil. This work was mainly done by women and children in the forestry communities in their home gardens. It involved much continuous work in a standing position. This process used a lot of fuelwood for boiling and the oil obtained was not clear, with a smoky taint. However, in Syria this was the only process in use, while in neighbouring countries, such as the Lebanon, a simple industrial process was used, with labour and fuelwood savings and a good quality final product. This process involved crushing the berries and extracting the oil with hexane and a rotary evaporator separator (Abi-Antoun et al. 2005).

Laurel oil processing in Tartous Province, Syria.

The oil was sold by the collectors for LS 250–500/kg. Trader mark-up was usually about LS 150/kg. For generations, about one-third of the total yearly income of laurel berry collectors in Kessab has come from this activity. Al-Hakim (1994) estimated that 50 t/yr of laurel oil was being produced in Kessab. The collectors sold the extracted oil to local soap makers and to the agents of soap factories in Aleppo, and to herbal shop traders, who came to the farms to purchase and then sold the oil in city markets in Aleppo and Damascus. Minimal intervention by the government, combined with poor infrastructure, limited information about chain actors, market information, etc., had resulted in an informal marketing system. The current poor market organization offered no incentives to improve the quality of production, or to improve sales channels. The production of oil was mainly at family level, and the technology used was primitive. The laurel berry harvest in Syria was very limited due to poor management of the resources, in part due to the existence of the regulation on the collection of forest products. Nevertheless, commercialization of laurel (berry) oil, in particular for soap making, was an important source of secondary and seasonal income for collectors, who could collect berries from wild trees without the need for any tools or inputs.

Laurel soap is believed to have been developed in Syria some 2000 years ago. In Syria, soap makers operated more than 50 small-scale factories (SEBC 2003), using very traditional technology, and selling under trademarks registered with the Ministry of Industry (Al-Hakim 1994). Most of the factories were concentrated in Aleppo Governorate, with a few in Kessab, the area where the laurel berries were collected. The soap-making process has for centuries remained unchanged. The soap was made with laurel oil, olive oil (first or second press), sometimes other vegetable oils, and caustic soda. The oil mix was stirred in large cauldrons with an aqueous solution containing the soda. This mixture was then heated to over 200°C with stirring until the oil fully converts, yielding glycerine and sodium salts. This process is called saponification. The remaining caustic soda solution was drained from the cauldron and the soap mixture left overnight to cool. Excess water was then drained off. Once a solid block was formed, the soap was cut manually into square bars, stamped, and stored in a dry place for at least six months. The process of making soap was carried out during the cold period, from November to April. It would not be possible to dry and cut the soap mixture during the hot period of the year, and for this reason work for soap-making employees was seasonal. From May to November, soap storage and selling were the main activities.

Organic soda was used up until the mid-1980s. It was obtained from the salsola plant (*Salsola kali* L.), a desert shrub that Bedouins still collect today from the wild, for a variety of traditional uses. Salsola plants were dried in the sun, then burnt and put in water for 6–7 days. The result of this process was then mixed with the other ingredients for the soap making. Organic soda has today largely been replaced by industrially produced soda. Most of the olive oil (approximately 80%) used in laurel soap manufacture was locally produced second quality oil. Laurel oil was mostly imported from Turkey (80%), as the local supply was insufficient for the needs of the industry.

The quality of the soap is determined by the quality and proportionate quantities of the oil and the other ingredients used in the soap-making process. Usually, the percentage of laurel oil in the oil mix ranged between 10% and 60%. The greater the amount of laurel oil, the higher the price of the soap to the final consumer, which varied between LS 80 and 300/kg. The margin for the laurel soap manufacturer was about LS 30/kg of sold soap. Laurel soap was very popular in Syria. Soap makers sold the soap to local consumers or to herbal shop retailers.

Laurel soap bar in the Aleppo souk, Syria.

In recent years, some soap producers in Aleppo and Kessab have developed a diversification of the product in terms of content and shape. This could be considered as the start of a marketing strategy. These new products have found a niche in the soap market and appealed to hotels, tourists, foreign traders and some local consumers. One factory in Kessab developed a high quality soap bar, smaller than the traditional shape, with a more sophisticated and colourful packaging. Product diversification was apparent from the creation of more than 10 different laurel oil soap bars, containing additional herbs with various properties (thymus for energizing, camomile for relaxing, *Pistacia palestinia* for scrubbing, etc.) and various shapes. This innovative laurel soap maintained the traditional recipe, but focused on a niche market since the price was more than LS 300/kg. These soap bars were only available at the herbal shop in Kessab belonging to the soap manufacturer in the area of the collection of laurel berries, or upon request to the small factory, which also exported a small quantity of product.

Another example of product diversification was found in a herbal shop of Aleppo market, where soap bars had been carved in the shape of Syrian monuments, for sale mainly to tourists. The price was about four times that of the traditional soap bars.

A small quantity of the soap produced was sold to foreign traders. Generally, the Syrian soap factories have not exported the product, but have sold ex-factory and the foreign traders had to bear all transport and customs costs. They then sold it to distributors that sold to herbal shops, mainly in Europe, the Gulf States, USA and Japan. For Europe, this transition, including the ex-factory cost, customs fees, transport costs, and trader, distributor and retail margins, increased the price to the final foreign consumers to about US$ 3–5/kg, depending on quality.

Quality

The quality of laurel oil differs according to the genetic character of the laurel used. Each variety is in fact characterized by a well-defined berry type, differing in scent, size, colour and fatty acid content. The extraction method also affects quality. The quality of the soap is also a factor of the other oils and ingredients used in the soap-making process (in terms of both quality and relative quantity). The greater the laurel oil quantity, the higher the price of the soap, as more laurel oil requires higher saponification temperatures and increased soda (Miller Cavitch 1997). For its natural and original processing, one of the oldest soap factories in Aleppo has received various quality awards from the Quality European Committee.

Market environment

In 1950, MAAR introduced a legal framework for the protection of natural forests in the country. These regulations restricted both harvests in the wild and cultivation of laurel trees and other species, including sumac, wild pistachio and carob. The 1994 regulations promulgated by MAAR replaced the 1950 legislation, and set out specific rules and regulations concerning civil responsibilities for the protection, investment and commercial use of all forest species, and consequent penalties for abuses committed. The legislation applied to state land, private land and protected areas. Utilization of the forest was subject to a licence, which was issued on presentation of all necessary documentation concerning the potential investor and those areas to be exploited. Any forest community member living within 5 km of the forestry areas had the right to collect forest products for household consumption only. This amount was determined by the forest capacity (e.g. 10 kg of laurel berries per collector in the Kessab area). Detailed regulations on the utilization of the forest were issued in the form of instructions and were distributed through various channels, including local farmers' cooperatives and Forestry Directorates. However, the rules were often inadequate or misunderstood by the forestry community members, who clearly were ill-informed about the regulation. Moreover, regulations might change periodically, and hence the regulations governing collection. For example, amendments to the 1994 forest legislation suspended any collection until 2006. Community members continued to harvest laurel trees in state lands and private lands according to community-based mechanisms, but very often technically in breach of national regulations. Rural communities should be encouraged to contact the authorities responsible to obtain more information and clarification regarding the regulations, in order to optimize the conservation and use of forest products.

The price of laurel oil and laurel soap was fixed by market forces. The price of laurel oil was sometimes subject to oligopoly. Labour and work-place conditions, with some exceptions for cleaner and more efficient soap factories, were quite poor for many seasonal workers, due to the uncomfortable positions (standing or sitting on the floor) and the high levels of dust and fumes coming from the saponification operation.

In the filière approach, the laurel production system can be considered a combination of 'market' and 'domestic' modes. The 'market' mode can be seen in the circulation of the product, which was mainly within a marketing channel, through commercial intermediaries, on the basis of a competitive price, and also in the nature of the dominant actors, being small-scale producers, intermediaries and retailers. The processing, using traditional rather than industrial techniques, corresponds to a 'domestic' mode.

Constraints and opportunities

Constraints

- There was a lack of awareness and concern amongst community members regarding the legislation regulating the utilization of forest products.
- The community structure dealing with the exploitation of laurel lacked any kind of organization.
- There was a lack of research on the laurel tree and on trade-offs between cultivation versus wild harvesting.
- There was an absence of methods to assess the quality of the oil produced. Only oil purchased from the Syrian farmers by the biggest soap factories was checked, but there was no standardization. The oil produced in Syria and the oil imported from Turkey were mixed together, with no opportunity to evaluate their respective quality or any differences.
- Support for efficient market strategies was lacking and there was little understanding of consumer or importer requirements. Problems included poor packaging, no quality control, no content indication, no factory trade mark for the international market, and almost no local niche market soap retailers.
- Funding and expertise were needed in order to develop market promotion on a larger scale, including advertising, Web sites, and appropriate presentation of product to foreign traders.
- Both the local and international market systems for the soap lacked any flow of information.
- Laurel leaf and berry collectors had difficulty gaining access to the market and obtaining early price information.
- There had been almost no investment in infrastructure for laurel oil extraction and soap making, and there was a need to improve labour conditions.
- Capacity building was lacking to develop an economic structure based on this resource, and no ability to assess the natural and human potential.

Opportunities

- Calls for better regulatory systems have already been voiced by local community members and soap manufacturers. However, there should be more involvement of forestry community members in the policy planning process. Among the possible alternative to this framework is the leasing to community members of laurel forest areas. Many community members and laurel soap producers feel that such a system would translate into a concrete incentive to promote the sustainable economic exploitation of local resources while maintaining an important natural resource for local development.
- Research opportunities exist in potential facilities and state nurseries already in place. The networks between nurseries and the private sector could be increased, as there is interest being shown by various stakeholders.
- With the increase in demand for export comes the need for greater attention to traditional processing, i.e. creating a protected traditional production process; improving quality control; and better marketing, including stamping the Syrian factory trademark on the soap.
- Possibility of entering a 'fair trade' market channel in Europe should be considered. Fair trade contacts could be established and the product adjusted to suit the market, such as adopting suitable packaging, supplying information about the product, etc. This opportunity could increase market benefit at the production level, but might imply the introduction of better labour conditions. At present, only the foreign traders are benefiting from the export markets for laurel soap.
- Follow-up the local niche market demand, with product diversification (retail points in Aleppo, Damascus, Lattakia and Palmyra).
- There are good labour opportunities in a healthy environment for the young generation that should be fostered, whilst maintaining the traditions.
- Capacity building of local human resources in technology and marketing could be implemented.
- Develop in-country capacity to extract essential oil from the leaves, and thus respond to and profit from the increasing demand from Europe.

CAPER

Mapping the value chain

The collection of caper flower buds was started only recently in Syria (in the 1990s) by nomadic rural communities, collecting in the summer period (June to August) and representing a secondary source of income. Bedouin, and other nomadic groups, harvested the flower buds for the Turkish food industry. Collectors were usually women and 7- to 14-year-old children. Each child was able to collect between 0.5 and 2 kg of caper flower buds per day.

Young caper collectors in Aleppo Province, Syria.

One individual was responsible for each group of collectors in each area. Those individuals liaised with the manager of a private enterprise, who provided them with material for collecting the capers (plastic boxes) and paid them about LS 65/kg of collected buds, according to the market price. Part of the money was used to pay the collectors (about LS 35/kg) and the rent of the premises where the first processing took place. Processing the caper buds involved entire families and consisted of sorting the caper buds by size, and then storing them in brine (16% salt solution) in plastic containers. The caper buds could be conserved like that for up to one year. These caper buds were then collected and stored by the private enterprise, who sold the capers to Turkish traders for about US$ 2/kg.

The Turkish factories transferred the capers to glass jars, adding preservatives and vinegar. The finished product was then labelled and sold to the European market (mostly Germany), for about US$ 2 per 75-g jar. All these steps in processing the product—from collection to distribution—made the mark-up very high. The flow of the caper buds through the value chain is shown in Figure 6.

There are no reliable figures available for the quantity of caper buds currently harvested in Syria. A realistic figure for the production of wild capers in Syria is approximately 4000 t/yr, which is about one-fourth of the potential caper harvest. The amount of capers collected, and hence the number of people involved in the collection operation, has depended on the annual Turkish demand, which has fluctuated greatly. Collection for Syrian consumption has been almost non-existent.

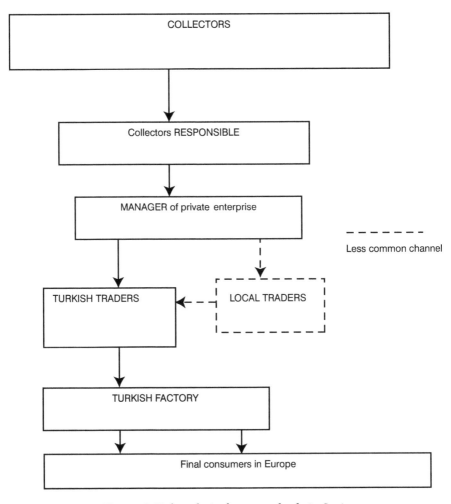

Figure 6. *Value chain for caper buds in Syria.*

Quality

The quality of caper buds was not checked by the collectors. During the very first handling by the collectors in their homes, capers were sorted using a very rudimental wooden tool with a metal net. Since the collectors were not trained to harvest the capers and they did not know the use of the caper buds by the final consumers, the bud quality was often low (variable size; many buds already flowering, hence inedible). However, the Turkish traders made a selection on the basis of quality and size of the product that they bought from the middlemen.

Market environment

National Law no. 7, 1994, concerning the protection and preservation of forestry areas (enacted by MAAR) replaced an earlier law from 1953. This law classified the caper shrub as a wild forest (*herage*) species. The law set specific rules and regulations, civil responsibilities, and penalties for improper use, for the protection, investment and commercial use of all forest species. The law applied to all land: state, private and protected areas. Utilization of and trade in the caper shrub was subject to a licence issued upon presentation of all the necessary documents concerning both the potential investor and those areas to be exploited. Prices were set by the market according to demand, which was strongly influenced by the level of demand from Turkish traders. In recent years, caper collectors have seen a decreasing trend in prices paid by the traders. In some remote, dry areas of Syria, namely Jabal-al-Hoss, collectors claim a collapse of the caper market in their area, since traders are no longer coming to purchase the product. Collectors are not aware of any reason for the collapse of the market, and they would be willing to collect capers for a price that would be even lower than in previous years.

Caper product display in Jabal-al-Hoss, Aleppo Province, Syria.

From the filière approach viewpoint, the caper production system can be considered to combine elements of both 'domestic' and 'market' modes, as the exchange was mainly done within a marketing channel, through commercial intermediaries, on the basis of a competitive price. However, the dominant actors were collectors belonging to fragile family categories (women and children) and the techniques used were very rudimentary and operated at household level.

Constraints and opportunities

Constraints

- There was a great lack of awareness amongst collectors about market channels and the final use of capers.
- There was no access to market information for collectors.
- There was no form of organized community structure dealing with the collection, processing and trade of caper buds.
- There was a lack of research on caper cultivation and on trade-offs for cultivation versus wild harvesting.
- Working conditions were very difficult during collection, due to the presence of thorns and the fact that the shrubs were either scattered across a large area or along the verges of highways.
- There was no quality control methodology operating at community level. Only the Turkish traders checked the quality of the capers, and hence they had the power to force down prices.
- There was a lack of support for efficient processing and storage.
- Capacity building to develop an economic structure based on this resource was lacking, as was any ability to assess the natural and human potential.
- There was a lack of expertise and training to improve collection, processing technology and marketing of the caper, and a lack of funding for investment in processing and storage infrastructure.
- The harvesting method of the wild caper was often unsustainable, and therefore represented a threat for this wild species.

Opportunities

- Collectors (ideally collectors *and* traders) could be organized in small companies.
- Collector cooperatives could be set up, with better access to the market and with the possibility of selling capers either direct to consumers or to foreign importers.
- The cultivation of capers could be introduced, but under experienced supervision to avoid the possible negative effects of introducing a cultivated species where only wild forms existed. Seed could be made available, pref-

erably of a thorn-less cultivar. In this way, caper production could increase and difficulties in collecting would decrease.

- Investigate the possibility of joint ventures with caper factories in other countries (e.g. Sicily, Italy) where the technology and the market are very well developed (see Annex 4). In response to a shortage of local labour, such factories would welcome joint-venture agreements with small farms in Syria.
- The culinary use of caper buds and caper fruits in Syrian cuisine should be promoted, and also in the households of caper collectors. This would raise awareness of the flavour of caper buds and fruit, the nutrients they contain, as well as the beneficial effects of this species.
- Develop a more extensive market for herbal remedy products based on the caper plant.
- Investment by government and international organizations could help facilitate the opening of small caper-processing factories for salting and glass-jar bottling of final product. The factories would involve local farmers and nomadic communities, who could be trained in processing.
- Women's groups in particular would benefit from any improvements in activities relating to caper collection and processing, due to the high percentage of women involved.

Box 5. Successful experiences abroad: capers on the island of Pantelleria, Italy

The quality of the product is determined by the techniques applied after collecting the buds. The caper cultivar used has been developed by farmers' cooperatives over the last 200 years. Once the caper is collected, it is carefully cleaned of any leaves or sand and is sorted according to size. The capers are then placed in terracotta containers, alternating layers of capers and sea salt. The use of solely sea salt preserves the strong aroma, the intense floral flavours and the firm texture of the delicate buds, which would be masked by vinegar or other preservatives. This process is repeated daily to aid lactic fermentation, and any liquid is drained off. The capers acquire the characteristics necessary for consumption after approximately ten days of treatment. Finally, they are bottled with sea salt and are ready for sale.

The collectors of Pantelleria formed a cooperative, producing, processing and marketing the capers. The vertical integration developed by these producers made possible a fair share of income among the chain actors, all members of the cooperative.

The cooperative has selected a high quality cultivar, Capparis spinosa *L.* var. inermis *cv.* Nocellara, *and obtained national and EU certification for "Protected Geographic Indication of Origin", increasing the value add-on of the product, which is sold direct by the cooperative to national and international markets and can be recognized by the consumers.*

Machine to pack capers in Pantelleria, Italy.

PURSLANE

Mapping the value chain

From April to December, purslane (*Portulaca oleracea* L.) was cultivated out-doors. During January–March, purslane was cultivated in greenhouses, mainly in the Damascus area, and then transported by farmers to other markets. Purslane was commonly sold on main, sub-main and local markets all over Syria. Usually, the traders selling purslane also sold other vegetables, in particular mint and parsley. Sometimes, farmers growing purslane also become traders and establish their own business. On the main market, wholesalers bought the purslane from the transporters (usually the farmers growing it) for about LS 10/kg and sold it to consumers for LS 15/kg, for a gain of LS 4–5/kg at most. Retailers also bought it from transporters for a maximum of LS 15/kg and sold it to consumers in their local markets for ca LS 20/kg (Figure 7). Purslane was sold to final consumers in packages of 200 g.

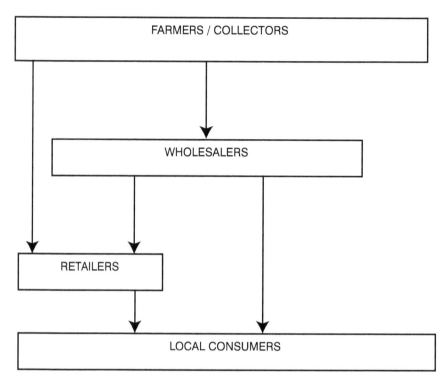

Figure 7. *Value chain for purslane in Syria.*

Purslane in the Aleppo market, Syria.

During the winter months, when purslane was cultivated in greenhouses, and during the month of Ramadan, when demand becomes very high, the price for consumers can reach LS 25/kg or more. Ramadan is the ninth month of the Muslim calendar, during which the Fast of Ramadan is observed. Lasting for the entire month, Muslims fast during the daylight hours and in the evening eat meals with friends and family. In some very poor markets in the main cities, purslane was sold for LS 10–12.5/kg. These markets offered very few vegetable species, but among which were wild and cultivated purslane. More than half of the supply arriving at the main wholesale market in Aleppo was sold on the same day. Because of its short shelf life, purslane could not be marketed the next day. Usually, traders and middlemen gave unsold purslane to farmers for use as small-ruminant feed. On the sub-main markets, it was not unusual to find some very poor people gathering the purslane and other vegetables left abandoned by some middlemen and selling them at a lower price.

Purslane in the Aleppo market, Syria.

Purslane seed was produced mainly in the Damascus area, which was considered the best area for this product. Each year, about 200 kg of seed was sold to farmers by each of the ten seed shops in Aleppo that sold purslane seed. Farmers generally bought about 500 g/yr seed, at a retail price of about LS 125/kg, compared with the traders buying-in at about LS 100/kg.

Quality

Cultivated purslane varieties had larger and more tender leaves, and matched local market consumer preferences better than the wild varieties. Wild purslane was in demand as a substitute for cultivated product when there was a supply shortage. Purslane was sold in the market as small bunches packed in plastic film or paper, and it was kept dry to keep it fresh. The only processing was trimming off of the roots. Each package was 200 g.

Market environment

The price was determined by the open market. Demand and supply balanced during the warm months, while demand exceeds supply during the cold months, leading to a higher price. During the month of Ramadan, when purslane salad (*fattoush*) is traditionally and commonly eaten, the market price almost doubled.

In the filière approach, the purslane production system can be considered as following a 'market' mode overall, as the exchange was mainly done within a marketing channel, through commercial intermediaries, on the basis of a competitive price. The dominant actors are small-scale farmers, intermediaries, wholesalers and retailers. However, the market deriving from the selling of purslane collected from the wild can certainly be considered a 'domestic' system, since the dominant actors involved in the filière are family members, in particular coming from fragile categories (elders, children).

Constraints and opportunities

Constraints

The primary problem in selling purslane was its short shelf life. In fact, fresh purslane leaves lasted at most half a day at room temperature (depending on the season) and about 4 days in the refrigerator (at about 4°C), before spoiling. This implies that all the supply should be sold to the consumers on the same day as arrival at the market, if refrigeration were not available. Production of purslane during the cold months (January–March) was possible only in greenhouses, and only big farms could afford to have one.

Opportunities

- To avoid conservation problems, purslane leaves should be packed in a manner that extends the shelf life. Purslane should be sold locally to avoid time lost in transportation. Refrigeration for transportation and storage of purslane should be available.
- Cultivation in greenhouses should be extended to areas other than Damascus. Techniques and greenhouse structures should be made available to farmers in other country areas.

- The use of purslane should be promoted in the cities and the countryside, to increase its presence in the diet of local people, and consequently to increase consumer demand.

MALLOW

Mapping the value chain

Mallow could be collected between February and March, with about three harvests possible, after which the plant was no longer good to eat. The market chain for mallow was quite simple (Figure 8). Since mallow was dispersed, farmers tended to collect it together with other plants, and sold in the market as a mix of edible plants (typically with *Brassica arvensis*, *Taraxacum officinale* and other herbs). Collection was done at family level, mainly by women and children. They were able to collect about 1.25 to 1.5 kg/hr, with an average collection of 7–10 kg in about 8 hours. Farmers collecting or cultivating mallow in the country sold direct to the final consumers in the market place (generally in the closest village or city). In bigger towns (like Aleppo), some farmers sold the collected mallow at the main wholesale market to retailers from smaller markets. The price of collected mallow was higher than the price of cultivated mallow. The wild mallow (collected) was sold in the market to final consumers for LS 30/kg, while cultivated mallow fetched less. The price of a collected mix of plants varied between LS 20/kg initially and LS 15/kg at the end of the day. Since the price in the main wholesale market was lower than the price set by retailers (about half the price), about 90% of the farmers selling in Aleppo tried to sell their mallow direct to final consumers in the markets. This implies that they would spend their day at the market. In the sub-main markets, it was not unusual to find some very poor people gathering the mallow and other vegetables abandoned by some middlemen and selling it at a lower price. Farmers kept some of the harvest for their own consumption. Consumer demand for fresh mallow has increased in recent years following the discovery of its beneficial health effects. Farmers cultivating mallow bought their seed at the seed shops in town (Aleppo), where it sold for about LS 100/kg. Each shop sold an average of 500 kg of mallow seeds per year. In recent years, farmers were typically buying about 25 g of seed to cultivate mallow for their own use. The demand for seed has increased as more farmers have started to cultivate mallow for the market.

Quality

In the market, consumers preferred the fresh wild mallow forms for their smaller leaves and better taste compared with the cultivated ones.

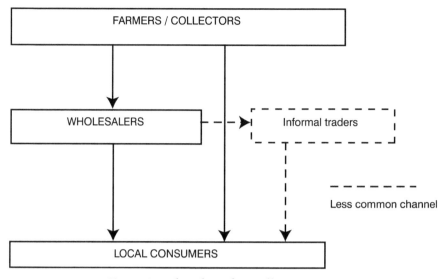

Figure 8. *Value chain for mallow in Syria.*

Market environment

The price was determined by the open market. Demand and supply were in balance. Seed demand and supply has been in equilibrium.

In the filière approach, mallow production could be considered a 'domestic' mode, as the actors involved in the chain belong to family lineages, and fragile groups in particular. Family members from rural communities are in charge of the production and commercialization. The exchanges are done at very informal level, including on a household exchange basis.

Constraints and opportunities

Constraints

- The very short shelf life was a major constraint for commercialization of fresh mallow leaves. The taste changes if the leaves are not eaten soon after being collected.
- Mallow was still cultivated on only a very small scale. There was a lack of cultivation research on this plant and very little specific information about mallow was available.
- With increasing mechanization in arable agriculture, the mallow that formerly grew on manually cultivated fields was destroyed. Hence, the collection of wild forms now implied additional labour, as the mallow grew scattered on uncultivated land.

- Sometimes, the harvesting of wild varieties was done in an unsustainable way (cutting off at the root), impeding regeneration.
- The collection and commercialization of mallow was looked upon as a 'business for the poorest'.
- The use of mallow in traditional cuisine was slowly disappearing among those living in towns, and among the younger generation.

Mallow stew in a household in Jabal-al-Hoss, Aleppo Province, Syria.

Opportunities

- Mallow is a very easily grown plant, succeeding in ordinary garden soil and in poor soils, without much care. Moreover, plants seem to be immune to rabbit predation.
- Because of its content, mallow is a very nutritional vegetable source, widely and cheaply available. The use of mallow should be promoted in the cities and the countryside to increase its use in local diets.
- There are some potential uses that are insufficiently exploited in Syria. The plant and the seed heads can provide yellow and green dyes; and a fibre can be obtained from the stems and used for cordage, textiles and paper making. Mallow can also be used as an ornamental flower plant.

- There is a commercialization potential in dried leaves for the emerging herbal product factories. Suppliers would be farmers and collectors from the wild in northern and central areas of Syria.

Box 6. Interesting case: herbal product factory in Aleppo

A factory was established in 2003 in Aleppo, with the function of processing and trading useful herbs for the market. At the time of writing, about 70 species were being processed, both cultivated and wild. Ten people were employed with various functions, including one woman. The suppliers were the farmers and collectors of wild herbs in northern and central Syria, and the buyers were distributors and final consumers. Distributors bought product for about LS 400 000/year. The products and the factory were inspected annually by the Ministry of Health, following the initial check before issuing the licence. In the factory, the herbs were tested, dried and stored.

Buying-in costs for the factory were about LS 70/kg for herbs, with LS 20/kg for processing, LS 20/kg for packaging and LS 30/kg for labour and machinery costs. Distributors purchased at LS 70 per 100-g package of finished product, and sold it to shops for LS 80 per package. The final price to consumers was LS 100–120 per package. Prices were fairly similar for most of the herbs.

AN OVERALL VIEW OF CONSTRAINTS AND OPPORTUNITIES

The constraints and opportunities for the six species are summarized in a matrix in Table 8.

Table 8. *Summary matrix of constraints and opportunities by species in Syria.*

Species	Constraints	Opportunities
Fig	Short shelf life as fresh fruit Storage problems for dried fruits (fungi and insects) Unattractive appearance of dried fruits Lack of knowledge of processing for quality Lack of market transparency Lack of price control for producers Lack of subsidies Lack of technical support	Biological production Better product quality Conservation of biodiversity Provision of technical support for cultivation to farmers Producer organizations Training in processing practices
Jujube	Very low local consumption Lack of awareness about the product Great fluctuation in prices (Lebanese taxes) Poor packaging and storage No quality control Lack of research on cultivation Lack of value-adding processing	Jam for local and export market Suitable for the local taste (similar to dates) Dried fruits can be dried *in situ* on the tree Training in processing practices
Laurel	Lack of awareness about the regulatory environment No producers' organizations Lack of research on laurel cultivation No quality control of the product No price control by the producers No processing techniques to ensure good quality Little understanding of consumer and importer requirements Poor packaging Lack of market transparency Lack of investment in infrastructure Poor working conditions	Better regulatory systems for natural resource exploitation Research on cultivation practices Improved quality in processing Exploit the export market (also fair trade) Product diversification Good labour opportunities Exploit the emerging foreign market for essential oil
Caper	Lack of market awareness and transparency for collectors No access to market for collectors No producers' organizations Lack of research on caper cultivation Collecting is hard work (heat and spines) No quality control of the product No price control by the producers Lack of value-added processing No good quality storage structures Unsustainable harvesting	Collector organizations Vertical integration of market actors Training in processing practices Research on cultivation practices Possible joint-ventures with caper factories in other countries Very nutritional source widely and cheaply available Promotion of its use in local diets Investment by government and development agencies in this market
Purslane	Very short shelf life No good packaging and storage Small-scale producers lack access to greenhouses for production during cold months	Improve quality of packaging Refrigerated transport and storage Cultivation in greenhouses Promotion of its use in local diets
Mallow	Very short shelf life Lack of cultivation and research on the plant Collecting it is hard work (scattered) Unsustainable harvesting "Business for the poorest" Traditional cuisine use is disappearing	Plant easily grown Very nutritional source, widely and cheaply available Promotion of its use in local diets Potential use as dyes for textiles and as paper fibre Dried leaves for herbal product factories

6

Findings: species and livelihoods

This chapter reports the results of the household field survey conducted between June 2003 and December 2004, working with the actors involved in the market chain of the target species. The analysis of the data was carried out at stakeholder level, involving collectors, growers, processors and traders. The findings have been organized based on the Sustainable Livelihoods Framework asset pentagon (See Annex 1), and focus on human and social capital, physical capital, financial capital and natural capital.

The data collected through interviews with households have been analysed at the chain actor level and again are reported using the Sustainable Livelihoods Framework. In the Sustainable Livelihoods Framework, the starting point focuses mainly on human and social assets, since they significantly affect NUS-related activities. Poor local communities often do not have the possibility of reaching or buying commodity crops, but have easier access to NUS. Moreover, the NUS-related activities are often off-farm, supporting a survival strategy.

The chapter ends with the analysis of the objectives of the actors in the market chain, the impact of commercializing diverse varieties on chain actor income share, and returns to labour. It concludes with the constraints perceived by the chain actors.

ORGANIZATION OF THE HOUSEHOLD FIELD SURVEY RESULTS

Objectives of the chain actors associated with the species used in the study are reported at the beginning to provide an overview of whom these actors are and why they are involved in these activities. The data on financial capital come from a simple self-assessment by the chain actors regarding their wealth and financial situation, including loans. The data reported in the human and social assets section cover issues related to household and family size, education, labour, gender, work experience, training and legal issues. The two asset elements have been combined since it was difficult to separate some issues—such as training and market relations—which, being very informal, stood between human and social capital. Physical capital includes data from farm size, land quality and irrigation on private land, and infrastructure. Natural capital is expressed by data on the perception of the chain actors regarding the land quality of the state and communal lands that they harvest, and on the availability of the species in terms of quality and quantity. Finally, data on seed availability are reported, followed by a section reporting data on income shares, and time dedicated to labour related to these activities. Data on market environment include price formation and time, and modes of reaching the market place. Data on the impact of commercializing different varieties have been reported, together with constraints on the development of commercial activities at various market-chain actor levels.

OBJECTIVES OF THE ACTORS IN THE MARKET CHAIN

Income generation was overall the main motive for the actors in the chain to be engaged in their activities. Other reasons reported were domestic use, nutrition and medicinal purposes, tradition, imitation of neighbours, ease of activity and diversification (Table 9). For collectors, income generation was the primary purpose, combined with domestic use. Frequency of replies noting income generation were slightly higher than those for nutrition and medicinal purposes. Tradition appeared as a third reason. For growers, the second reason was tradition, followed by nutrition and medicinal use. Only a few growers replied that they carried out this activity to imitate their neighbours or because of the great adaptability of the species to the land. As expected, reasons such as nutrition, tradition and adaptability of the plant appeared more often in conjunction with a small or marginal cultivation area, while income was the main, and often the only, purpose of growers with bigger parcels of cultivated land (mainly for fig and purslane).

For processors and traders, 'income' was the main motive, followed by nutrition and medicinal use purposes. For processors, tradition and imitation of neighbours came as minor reasons. Many traders started their business with these species because of tradition or to imitate neighbours. A few traders specified 'diversification' as another reason to commercialize these products.

Regarding destination of the product, Figure 9 shows that the sales share was very important for all the actors in the chain, in particular for traders (95%), growers (86%) and processors (80%). This data confirms the key role that markets play in sustainable livelihoods related to NUS.

Household consumption (mainly for nutrition, medicinal use and animal feed) was quite significant among collectors (36%). Processors (21%), more than growers (14%), utilize the product for household consumption. This can be explained by the fact that some of the processors interviewed were growers that processed any remaining product for household consumption or for sale as a secondary product.

Table 9. *Relative frequencies of motives for participation in particular market-chain activities as reported by the actors.*

Chain actors	Income	Nutrition/ medicinal	Tradition	Imitation of neighbour	Ease	Diversification
Collectors	«««««	«««««	««			
Growers	«««««	«««	««	««	««	
Processors	«««««	«««	««	««		
Traders	«««««	«««	««	««		««

Notes: Respondents could reply with two reasons; more « indicates greater frequencies reported by chain actors.

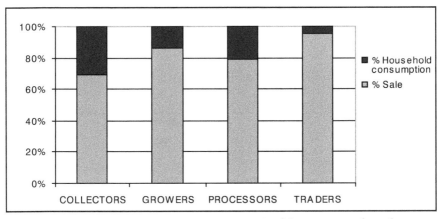

Figure 9. *Average proportion (percentage) of destination of product per chain actor.*

FINANCIAL ASSETS (WEALTH)

To have an overview of the financial assets held by the different chain actors, a self-assessment was performed, based on the indicators shown in Table 10.

The results of this self-assessment when converted to a percentage basis (Table 11) show that collectors had the greatest percentage of poor households (57%), while most of the growers (67%) and processors (61%) lived in intermediate wealth conditions. The greatest percentage of rich was among the traders (29%). This could be explained by the fact that product value addition was concentrated at the end of the market chain. However, there was a great disparity among the trader group, as poor households were still well represented (30%). This could be explained by the fact that many poor collectors (and processors and growers in a few cases) were also traders, as they dealt with final consumers in the city or village markets.

Loans

Respondents from collectors, processors and traders groups indicated that loans were difficult to obtain, if not unavailable, to start up or to carry out their activity in the market chain. Only two of the interviewed growers mentioned that they had received a loan to start up their activity, for the cultivation of mallow and purslane (about LS 25 000 from a government bank), while others received money from friends or relatives. As was quite common in Syria, the various chain actors might have obtained inputs from traders for repayment after the sale of the product, but with a price penalty (Buerli and Aw-Hassan 2004).

Table 10. *Indicators for an informal wealth self-assessment.*

	Poor	Average	Well-off
Own land	1–3 ha	15–20 ha	Large area
Own sheep	0–5 head	10–50 head	>50 head
Other livestock			Yes
Own a TV		Yes	Yes
Own a radio	Yes	Yes	Yes
Own a vehicle			Yes
Food for consumption	some months with food difficulties	meals with meat for some months	meals with meat all year
Water availability		Yes	Yes

Table 11. *Percentage distribution of actors in market chain by their own assessment of their wealth status.*

Chain actors	Poor	Average	Well-off
Collectors	58	41	2
Growers	25	67	8
Processors	29	61	11
Traders	30	41	29

HUMAN AND SOCIAL ASSETS

Family size

For this study, family was defined as the unit formed by parents and their children. 'Parents' were usually a married couple, but sometimes a widow(er) or a man with two wives. The average family size of the households interviewed was 7.54 family members. As is shown in Table 12, neither average family size nor the age of the household head presented a wide range of values, with no difference between rural and urban areas.

The average age of the household head varied between 46.25 years for growers and 49.82 years for traders.

Table 12. *Demographic household characteristics of the various actors.*

Chain actor	Average family size	Age of the household head
Collectors	7.21	49.60
Growers	7.34	46.25
Processors	8.28	48.34
Traders	7.33	49.82

Education

For education, the lowest level was found among the collectors, highlighting a correlation with less wealth. The proportion of the collectors that were illiterate was 64%, compared with illiteracy levels of 53% among growers, 18% among processors and 28% among traders (Figure 10). The link between education and wealth was also substantially confirmed for other groups.

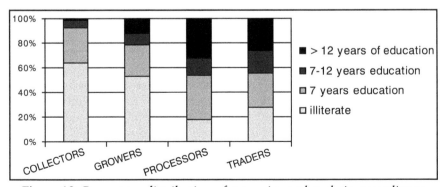

Figure 10. *Percentage distribution of actors in market chain according to level of education.*

Labour issues

Table 13 shows the average labour share for market actors in the chain. Labour share is defined as the percentage of their total working that an actor devotes to that specific activity in a year. The average annual labour share for the species under study was 20% for growers, 13% for collectors, 17% for processors and 18% for traders. In particular, caper and laurel collectors reported the highest average percentage among the collectors (up to 30%). A possible justification is that laurel collection is very well remunerated. Caper collection was performed by poor nomad groups (Bedouins) living in remote desert areas, who relied very much on this summer income-generating activity, although the remuneration was quite low. Among processors, also, there were great differences between the labour share of fig and caper processors (ca 9% each) and the 60% labour share of laurel processors. Processors of laurel were generally soap makers, who dedicated all their time to this activity.

Table 13. *Labour share for actors in the market chain.*

Chain actor	Labour share (%)
Growers	20
Collectors	13
Processors	17
Traders	18

Table 14. *Average numbers of total workers and family members, and average ratio of family to total workers, for market-chain activities.*

Chain actors	Average total number of workers	Average number of family members	Proportion of family in total workers (%)
Collectors	5.26	5.12	98
Growers	6.15	4.15	67
Processor	6.61	3.94	60
Traders	3.00	2.34	78

Table 14 shows that growing as an activity implied the greatest numbers of workers (6.15), and that 68% of the workers were family members, while the other 30% were paid external workers. Collecting implies on average 5.26 people, and almost all (97%) of the workers involved were family members. This can be explained by the fact that collecting was the most unstructured activity, which was carried out without requiring funds and facilities. Processing had quite a high labour demand, as it involved very traditional processing methods, carried out mostly by family members (60% on average). There was a relative low average ratio of family members to workers in small-scale laurel soap factories, where there was a high average ratio of external workers employed. Trading was not a very demanding activity in terms of labour (3 people on average), and the average ratio of family members involved in the activity was quite high (about 78%).

Gender

Traditionally in Syria, women are very much involved and in charge of cultivation and collection activities, while men are more dedicated to trading. The survey showed a significant presence of women in market-chain activities for the species studied, in particular at the start of the chain: collection and growing activities. Children (under the age of 12) also often took part in chain activities, in particular for seasonal work. Table 15 shows, for the households interviewed, the average number of women and children involved in the market-chain activities of the respective species, and the proportion of women and children involved in the activities. Women and children are more obvious in collection activities, with a ratio of 53% and 29%, respectively. Growing as an activity also sees the involvement of a great number of women (37.81%), with help from the children (5% of all workers). Women are also greatly involved in post-harvesting and processing activities, at 34.31% of all labourers, while they are not very involved in trading, where more children (20.56%) than women (11.49%) were engaged in transporting and sales assistance.

Table 15. *Average numbers of total women workers, child workers and average proportions in all workers for the market-chain activity.*

Chain actor	Average total no. of workers	Average no. of women workers	Average proportion of women in total (%)	Average no. of child workers	Average proportion of children in total (%)
Collectors	5.26	2.77	52.59	1.53	30
Growers	6.15	2.33	37.81	0.30	5
Processors	6.61	2.27	34.31	0.00	0
Traders	3.00	0.35	11.49	0.62	21

Work experience

Despite the fact that most of the species studied and the activities around them are considered traditional, the mean years of experience of chain actors reported in Table 16 reveal that the chain actors had relatively short experience in the activity that their household carried out with these species, namely 16 years for collectors, 20 years for growers, 35 years for processors, and 21 years for traders. This result can be explained by the fact that some chain activities were relatively new for some species. For example, there were cases where laurel growers had only one or two years of experience, since, traditionally, laurel has only been harvested from the wild. Most of the caper collectors had ten years or less of experience, as did the caper processors. Although there is some evidence that caper has been used as a medicinal plant for many centuries, the caper bud business was very recent.

Caper collecting in Aleppo, Syria.

Table 16. *Mean of years of experience of chain actors, per species and mean for all species.*

Chain actor	Fig	Jujube	Laurel	Caper	Purslane	Mallow	Mean
Collectors	—	—	17	8	20	19	16
Growers	42	32	1	—	18	7	20
Processors	37	—	55	12	—	—	35
Traders	22	23	24	22	15	19	21

Nevertheless, there are great differences in terms of years of experience when the various species are considered. For example, some fig growers interviewed declared family experience in the activity of about 100 years. Looking at human capital, the means of family size, relative extent of education and average years of experience per chain actor (Table 17), it seems apparent that years of experience and education have grown in parallel. There was little influence of family size on education and years of experience.

Table 17. *Human capital parameters and percentage distributions among the chain actors.*

Human capital indicator	Collectors	Growers	Processors	Traders
Family size	7.21	7.34	8.28	7.33
Education (%)	36	47	82	72
Years of experience	16	20	35	21

In informal discussions, chain actors described the process that led them from one activity to another. For example, some growers became traders; and some collectors became processors. This process can be seen as a 'maturation' that moves the household activity from one activity to the next one in the chain. This generates a slow *vertical integration* in household activities, such that the households dedicated to cultivation tend to become processors or traders. There was no sign of development of *horizontal integration*. Cooperative efforts in a commercial sense were almost inexistent.

Training

Among growers, processors and traders, training was informal within the households' members (Table 18). More than training, it was a transfer of experience and knowledge between generations. Some growers indicated that some workers, including a few women, had received training from other community farmers. For processors of laurel products, the factory owners were responsible for training their workers. Traders also indicated some training from members of other communities that trade in the same market place. No form of training was reported by collectors.

Table 18. *Training source and common training needs identified by the actors.*

Chain actors	Source of training	Training needs
Collectors	No training	——
Growers	Household, community farmers	Storage, packaging, cultivation practices, manufacturing
Processors	Household, factory owner	Improved processing methods
Traders	Household, members from other communities	Packaging, marketing

Regarding training needs, to improve activity output, growers repeatedly expressed their need for training in storage, packaging and manufacturing. The training needs defined by the chain actors indicate an inclination towards vertical integration of their activities in the chain, including cultivating and processing. A few respondents added cultivation practices to the list of training needs, in particular for fig and jujube trees. Traders showed interest in training for packaging and better marketing practices. Though the collectors usually receive no form of formal or informal training, they did not express any need to develop a particular skill or develop their knowledge. However, many of them wanted to know how they could derive more income from their collection activity.

Legal issues

Regulations introduced by MAAR affected the collection, processing and commercialization of products derived from wild forest species. Most of those living in the forestry areas were aware of the regulations, but there was confusion among community members concerning the application of those rules. In some forestry communities, such as Kessab (Lattakia Govenorate) and Kadmus (Tartus Govenorate), there were informal rules regulating the collection of wild species from public land. These were rules and behaviour implicitly imposed by the families of the communities. Each family or group of families had access to plants growing in particular areas of public land. Community members were in close contact with forestry guards, but nevertheless the application of the government regulations differed from area to area.

PHYSICAL ASSETS

Farm size

From the tenure point of view, cultivable land was to a large extent private (98% in 1993), while the uncultivable land was both private and public, with a slight prevalence of the latter (Sarris 2001). Private land included cultivated land, under rain-fed or irrigated conditions, in addition to fallow and some uncultivated land. According to UNDP/GEF studies in 2000, the average land ownership per agricultural family was 9.6 ha in 1993, with ownership of holdings classified as small-scale (68.5%), medium-scale (30.6%) and large-scale (9.0%). In addition, there was a system of state land ownership, cooperative ownership and endowment ownership, distributed among small-scale (81%), medium-scale (18.3%) and large-scale (7.0%) landholders. The average size of holdings has been decreasing over time, but there was some discrepancy in the values reported in different sources, despite all being derived from the same census data (Sarris 2001).

The household field survey in this study highlighted the fact that growers usually cultivated the target species on private land, where they lived on their farms, and where they grew the species on a small scale. Of the growers interviewed, 96% owned the land, and only 4% rented, with such land leasing limited to only a few cases of major cultivation (figs and purslane), where the growers work the land of the owner and they receive a monthly salary in return, while all income coming from the sale of the harvest was kept by the landlord. The greatest percentages of cultivated land were for fig and purslane. On average, the total area cultivated with the targeted species was 0.44 ha (including two special cases of fig cultivation of 3 ha). As shown in Figure 11, compared with growers, fewer collectors owned their land, as 66% of the collectors interviewed owned some land, while the remaining 34% owned none.

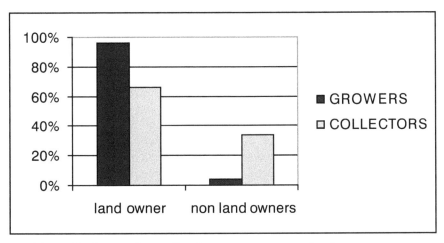

Figure 11. *Proportions of landowners versus non-landowners among growers and collectors.*

Land quality and irrigation

On a scale of 'bad, average, good, very good', 29.41% of the growers interviewed indicated that their land was of average quality, while 60.29% said that it was good and 10.29% considered their land to be very good. Regarding water regime, 48% of the farmers interviewed indicated that they irrigated when cultivating the species in this study, while 52% of the growers said that the species they cultivated were rain-fed only. However, Table 19 shows that only in the case of cultivation of vegetables was irrigation common, while fruit trees survive under rain-fed conditions. More than 86% of the cultivated figs were rain-fed, with a very few cases of irrigation with sewage water; more than 62% of the jujube trees did not need irrigation; and all the growers of laurel trees indicated that they did not water their plants.

Table 19. *Proportions of rain-fed and irrigated cultivation among the cultivated species.*

Cultivated species	Rain-fed	Irrigated
Fig	86%	14%
Jujube	63%	37%
Laurel	100%	0%
Purslane	0%	100%
Mallow	0%	100%

Technologies used

Among the growers interviewed, 50% indicated that they used pesticides and tractors, while 52% of them used fertilizers. Fertilizers and pesticides were used in both rain-fed and irrigated cultivation. The processing methods used by processors were all natural—sun drying, boiling, and addition of salt, sugar, vinegar or other herbs—with no use of modern technology and equipment. Processing was done on their own farm (53% of the processors interviewed), at other farms (31%), or at home (16%) (Table 20). These data show that processing was a secondary activity for the selected species. The exception was laurel processing (laurel oil and laurel soap production) as this activity required specific machines and equipment (though rudimentary), and dedicated workshops or factories. In most cases, processors of one species did not process other species, the exceptions being fig processors who also processed grapes and apple.

Table 20. *Site of processing (excluding laurel soap production factories).*

Home	Own farm	Others' farm
15%	54%	31%

NATURAL ASSETS

The target species considered in this study are cultivated or collected in various areas in Syria, from the arid to the humid zone. Collectors gather the wild species either from communal land or from forests that are mainly controlled by the state and denominated state land (Sarris 2001). Collectors did not know how many hectares they harvest, but they did have some idea of how many hours they had to walk to harvest the product.

Caper plants in Jabal Hoss, Aleppo Province, Syria.

Land quality

On a scale of 'bad, average, good, very good', 10% of the collectors interviewed perceived the lands where they harvested the wild species to be of bad quality. More than half of the respondents indicated that they considered the land to be of average quality, while about 38% of the collectors thought that they harvested from good quality lands. Only 2% considered the lands they harvested to be of very good quality. Comparing these data with the information gathered from the farmers (Table 21), it seems that the cultivated private land was more productive, or was considered to be of a better quality.

Table 21. *Perception of collectors and growers regarding the quality of the land harvested.*

Chain actor	Bad	Average	Good	Very good
Collectors	10	50	38	2
Growers	0	30	60	10

Availability of the species

More than half of the collectors interviewed indicated that the quantity and the quality of the wild species that they harvested were comparable to 10 years previously. However, more than 30% of the collectors thought that the species were becoming less available. More than 20% of the collectors stated that the quality of the species had improved over the previous decade, while 13% thought that the quality had decreased (Table 22). These data might indicate a negative impact from the harvesting of the wild species, which were slowly disappearing, probably due to unsustainable collection. The data on quality indicated an improvement in quality, possibly due to selectivity on the part of the collectors.

Table 22. *Perception of the collectors regarding the availability of the wild species in terms of quality and quantity, compared with a decade previously.*

	More	Similar	Less
Quality	24%	63%	13%
Quantity	0%	67%	33%

Seed availability

More than half of the growers (58.21%) obtained their seed from their own production. About 30% of the growers interviewed bought their seed from shops, while only 12% obtained seed from other farmers within or outside their community. This data (Figure 12) can be interpreted as lack of collaboration among farmers.

Figure 12. *Seed source for growers of the species in the study.*

INCOME SHARES AND
RETURNS TO LABOUR

Income shares

Income share is here defined as the portion of yearly income generated from the activities related to the specific species, out of the total annual income of the actor. Average income shares were calculated per chain actor. In the calculation, only respondents indicating an income share of more than 1% were included. In addition, the two laurel processors owning laurel soap factories, for whom the activity was their only income source (100%), were excluded.

Income share derived from the NUS-related activities varied from about 10% for processors, 11.25% for collectors, about 22% for traders, and to 23% for growers. Results show that for processors (excluding the laurel soap processors) and collectors, the activity was certainly more marginal than for the other actors in the chain. On average, the income generated by the commercialization of the target species represented between 10 and 20% of the yearly income of the actors in the market chain (Table 23), with a coefficient of variation varying between 0.53 and 0.87. These data confirm the strategic value of NUS.

Table 23. *Average share of species sales in total household income.*

Chain actor	Share in average income (%)	Coefficient of variation
Collectors	11	0.65
Growers	23	0.78
Processors	10	0.53
Traders	22	0.87

Table 24. *Use of the income for the chain actors generated from the activity (number of responses).*

Chain actors	Household needs	Nutrition	Education	Work
Collectors	14	10	6	—
Growers	37	30	16	6
Processor	5	5	5	1
Traders	8	5	5	5

Note: Respondent could give more than one use.

The earnings coming from the activities associated with the target NUS can be considered subsidiary in the overall national economy, but are quite significant for the households of those involved. In fact, this marginal and mostly seasonal revenue was considered good additional income by the rural community households involved in the chain activities. Chain actors indicated some priorities in using the cash generated by these activities. All the actors seemed as a priority to use the marginal income to cover household needs (Table 24).

The use of the income to cover some household nutrition expenses was reported more frequently by growers and collectors. Finally, all actors indicated the use of the income to cover education expenses. Only collectors seem not to use the cash generated by these activities to cover work-related expenses.

Returns to labour

Figure 13 shows income shares and returns to labour, comparing the income share averages by activity along the market chain, and the labour share in the various activities by the actors. From the collected data, growing and trading proved to be the most profitable activities, as the income shares—23% and 22%, respectively—were higher than labour shares (20% and 18%). Collection activity showed low remuneration, with an income share equal to 11% and labour share of 13%. As described in the labour section, this can be explained by the fact that collecting was a very labour-intensive activity, and it was characterized by lack of facilities and harsh conditions. Processing seems to have been the least remunerative activity (income share of 10% and labour share of 18%). This result ignores the income share equal to 100% reported by the two laurel soap processors. For the other species, processing (mainly cleaning and drying) was not yet a remunerative activity, as the processing did not bring any real value addition.

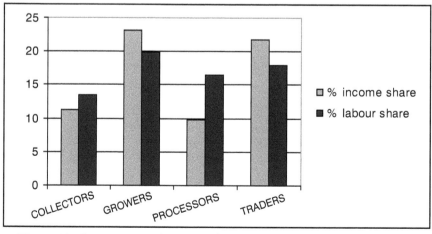

Figure 13. *Comparison of income shares and labour shares of activities, by actor in market chain.*

MARKET ENVIRONMENTS

Prices

The Syrian government has tried to develop the agricultural sector in particular and rural areas in general through programmes, plans for pricing, marketing and loans. Nevertheless, there was still no market transparency, and market intelligence was lacking. This was more relevant for minor species, where the market was more informal and not organized. Most of the chain actors interviewed indicated that the price of these products was set by the market (Table 25). For growers, the open market was the only price regulation system, while about half of the processors and collectors indicated that the price was determined by a monopolistic market, created by intermediaries. Collectors and traders acknowledged some government influence on prices.

Table 25. *Percentage of actors indicating a mode of price regulation for their products.*

Chain actor	By government	Market monopoly	Open market
Collectors	7	56	38
Growers	–	–	100
Processors	–	46	54
Traders	4	38	57

Regarding the demand:supply balance and product quality, processors indicated that the quality and the quantity of the product had diminished in the previous ten years. Traders indicated an opposite trend. This could be an indication that the market for these products was more suitable for raw products than for processed ones. On average, demand and offer seemed to be equal in all the groups.

Reaching the market place

In recent years, the government has provided a good level of specialized agricultural services and has established a basic structure that can facilitate development, such as roads, means of transport, energy and various services. These efforts have resulted in a relative improvement in standards of living for people living in the rural areas. This has had a positive effect on the production and commercialization of minor agriculture products. However, in the opinion of Syrian experts, this improvement has not reached the level required to ease the severity of poverty and reduce its social effects in the countryside.

From the survey, it seems that growers spent on average about LS 68 to reach the village market, and also the city main wholesale markets and city sub-main markets, with the average time being 46 minutes. About 20% of the collectors interviewed walked to the nearest village market, without spending money for transportation. The remaining 80% usually sold their product in the city sub-main market, spending on average LS 67 and travelling for an average of 64 minutes. Two-thirds of the processors interviewed used the city main wholesale markets to sell their product, while one-third sold in the village market. The average transportation cost for the processors was LS 43, for an average journey of less than an hour. Traders mainly sold through city shops and city main wholesale markets. They took about an hour to reach the marketplace, at a cost of LS 31.

Table 26 shows the various times and costs for the market-chain actors to reach the market place. Collectors and growers were more affected by transport cost and took longer. Areas exploited for the collection of wild species were usually further from the villages or cities where the products are sold, compared with farms. This might explain why—on average—collectors took more time to reach the market place. Traders were less affected by transportation cost and time, since they operated in a more restricted area. For example, they bought from the city main wholesale market and sold to retail shops in the same city. The cost for transport for farmers and traders of jujube to travel across the Lebanese border could be up to LS 2500–3000, with an average journey time of 6 to 8 hours. All the actors interviewed indicated that the market places were easily accessible to consumers.

Table 26. *Average cost and time to reach the market place for chain actors.*

Chain actors	Average cost to reach the market place	Average time to reach the market place
Growers	LS 68	46 minutes
Collectors	LS 67	64 minutes
Processors	LS 43	55 minutes
Traders	LS 31	53 minutes

Commercializing different varieties

For jujube, caper and mallow, chain actors did not distinguish different varieties. In the case of purslane, farmers and traders distinguished two varieties according to harvesting time: *rabieh* for the spring variety, and *hamga* for the summer variety. However, according to studies of the University of Aleppo, the genetic diversity was very low, and the chain actors did not mention any difference in cultivation or market between the two forms. In the case of laurel, collectors and some processors were aware that different varieties of laurel trees produce different berries with various characteristics that define the quality of the extracted oil, either for its use in cosmetics or in soap manufacture. However, there was no substantial differentiation among these varieties in marketing terms for collectors, growers, processors and traders. In the case of figs, the chain actors interviewed indicated that variations were recognized among the varieties in terms of both cultivation and trade. Table 27 relates the number of varieties cultivated and traded with the resultant average income share. Among the growers, 86% cultivated three or more fig varieties. From the data gathered, it was calculated that the increase in the income share of the growers depended on the increase in the number of varieties grown (Pearson correlation rate = 0.474; statistically significant at 99% of t-two-tailed-confidence interval). This indicates the value of intra-species diversification for the benefit of grower livelihoods. In contrast, the extent of intra-species diversification among traders and the trader's income (livelihood benefit) are negatively related, with a Pearson correlation rate of -0.241. The negative correlation rate among traders can be interpreted as a threat in terms of biodiversity conservation. Reduction in varietal diversification results in increasing specialization of the product available on the market. For processors, the use of different varieties was very weakly correlated with their income share (r=0.14).

More extensive analysis using genetic and economic analysis is needed to measure the impact of the conservation of biodiversity on the actors' income generation, and on their livelihoods.

Table 27. *Number of growers and traders cultivating a given number of fig varieties, and average income share.*

No. of varieties	No. of growers	Income share for growers (%)	No. of traders	Income share for traders (%)
1	1	1	10	35
2	4	14	6	23
3	25	16	3	23
4	2	24	1	20
5	2	30	2	20
6	2	30	2	18

CONSTRAINTS TO THE DEVELOPMENT OF COMMERCIAL ACTIVITIES

From a qualitative analysis of the research data and observations, it was found that many constraints hampered the development of the market potential of these species at various levels of the chain.
- At the level of cultivation, proper agronomic knowledge and application of techniques for the cultivation of most of the species were still lacking.

- At processor level, added-value processes complying with accepted quality standards were not available. While unique local knowledge existed on traditional uses for each species, the transfer of such knowledge from older community members to new generations and end users was very limited and the knowledge itself was fast disappearing.
- At collector level, the labour force was often formed from fragile social groups, e.g. young children, nomad women and elderly people.
- At trader level, marketing intelligence was greatly lacking. Information on demand and supply of products at both local and national level was largely unavailable or very limited.

The whole chain was characterized by a lack of any solid market structure and organization. Markets lacked any coordinated approach among their members, trustful relationships among actors, market transparency or vertical integration. Existing enterprises were few and working in isolation. Horizontal integration was also very weak: there was no cooperative effort among local growers and local traders. At policy level, there was a lack of clear and supportive policy frameworks to sustain and develop the market chain of these species. Table 28 reports the major constraints indicated by the chain actors interviewed for each species. Growers were mostly concerned about problems associated with cultivation practices. The main constraint for collectors was the labour-intense nature of their activity, due to the absolute lack of facilities, infrastructure and labour organization. The major constraint for traders was packaging, which was not suitable for the final users. As for indications of how to improve the quality of the products, packaging, appearance, taste and size were the key factors. The chain actors indicated some obvious improvements that they thought would boost their activities. Collectors and growers thought that they could increase demand for their products by improving the product's taste and appearance. Processors thought about applying different processing methods to add more value to the product to meet demand. They would like to find more profitable marketing channels through which they could market their product, since most of them did not trust the current traders and intermediaries. Traders thought that improving the packaging would be useful for improving the market for their products, and they called for improvements to boost demand (advertising, product awareness raising, etc.).

Table 28. *Main constraints identified by the chain actors and their proposed improvement to develop their activities.*

Chain actor	Constraints	Improvements proposed by chain actors
Collectors	Labour conditions and lack of infrastructure	Improve the appearance of the product
Growers	Cultivation practices, packaging	Improve the taste and appearance of the product
Processors	Labour conditions, demand constraints	Apply different processing methods, change marketing channel
Traders	Lack of suitable packaging	Improve the packaging, boost demand

7
Conclusions and recommendations

The research data confirm the strategic role played by these NUS in the livelihood strategies of rural households in Syria, and the importance of strengthening markets if these strategies are to be supported and these species preserved. Human and social capital were the main livelihood assets on which small-scale producers of NUS relied. NUS were very adaptable to the natural environment, but needed to be addressed by research on conservation and on sustainable use issues. However, farmers were willing to grow and collect these species in a sustainable manner only if they received benefits from them. These species are grown, collected, processed and marketed locally. Nevertheless, the lack of market organization was impeding the actors, in particular those at the top of the chain, from obtaining benefits. Therefore, interventions at production, marketing and policy levels to increase market potentials were needed. Capacity building and knowledge sharing concerning cultivation practices and marketing strategies, reorganization of the market relationships and private sector partnerships might help generate market value, serving the dual goals of supporting rural livelihoods and maintaining biodiversity. The detailed conclusions and implications reported below have been based on a combination of the results of the market-chain analysis and the livelihood household survey.

LIVELIHOOD OF THE MARKET-CHAIN ACTORS

This pilot study shows that the income share from NUS activities, while minor for the overall national economy, could be considered an important contribution to the livelihoods of rural community households engaging in these activities. The income they generated varied between 10 and 23% of annual household income. The very high proportions sold (64–95%) compared with kept for home use confirm the key role that markets play in the livelihood strategies related to NUS. Income generation was the primary purpose for engaging in these activities. Nevertheless, these species were also important for food security and health remedies, in particular for collectors, who kept an average of 36% of the product for household consumption (Figure 9). These results confirm the role that diversity of plant genetic resources represented by NUS played in livelihoods of all the market-chain actor groups. The collectors emerged as the most vulnerable chain actors group in terms of livelihood assets, with the highest rate of illiteracy (>60%) and with more than half of the households considering themselves poor. Illiteracy was also quite high among growers. The labour share of the market actors in the NUS activities was about 15–20% of total activities. Growers and traders dedicated more time to these activities. The highest income:labour share ratios were found for growing and trading activities, while collection and processing showed less income share relative to the amount of time required. Collecting was a very labour-intensive activity, and it was characterized by lack of facilities, such as appropriate collecting and sorting tools and storage rooms, as well as harsh working conditions.

All activities were carried out at household level, mainly by family members. In the case of collectors, the ratio of family workers to total workers was over 97%. A small proportion of external workers could be found among growers and processors. The involvement of women was very high, in particular for collection (>50%), growing and processing (ca 35%). Children below 12 years of age were often involved in collecting, growing and trading activities. From the work experience point of view, there were some differences among the species. Some activities could be considered new, such as the collection of and trade in capers, where the actors' work experience built on less than 10 years involvement. In contrast, chain actors working on 'traditional' species such as laurel and laurel products reported over a century of family experience.

The source of training for growers, processors and traders generally was from within the same household. Nevertheless, growers also received training from other community members, and traders from other communities. Processors of laurel soap were trained by factory owners. The only group not receiving training were the collectors in the wild. While the growers, processors and traders attested the need for relevant training in their respective fields, the collectors indicated no need for training. It was interesting to see the concern of the growers in manufacturing and packaging practices, showing a tendency toward vertical integration. Most of the actors in the chain initiated their activities on NUS for income reasons, with nutrition and medicinal needs as subsidiary reasons. Other reasons were tradition, imitating neighbours, the suitability of the plants (including their adaptability), and a willingness to diversify their crop base.

LOCAL USE AND MARKET CHAIN

The target species were called by the same names throughout the country, with the exception of fig varieties, which had different names in different areas. Some species have been used traditionally for household nutrition (fig, mallow, jujube and purslane) and consumption (laurel soap and leaves). Others (caper) were considered new crops and were not collected for household consumption, but only for the market, upon request from intermediaries. Some species were seen as 'food of the poor' (mallow, for example) and so both household consumption and trade were less attractive, particularly for people living in towns. Some species have been traditionally used as herbal remedies (mallow, jujube) by the rural communities that know their properties. Some others (caper, jujube) were sold in herbal shops and only herbalists knew their properties and could advise customers on their use. To developed increased yet sustainable use of these species, there is a need to raise awareness about traditional uses, nutritional properties and health benefits, among both rural and urban dwellers. For the species studied, as for NUS in general, the value chains were generally short and not well organized. According to the definition of the filière (Hugon 1995) typology of the production system modes, the chains regulating the species studied were either the domestic mode or the market mode, or a combination.

The trading system was simple and usually did not imply many intermediaries between the producers or collectors and the traders. The actors often had multiple roles: family small-scale farmers were simultaneously entrepreneurs, employers and labourers, as well as producers, processors and traders. The employers often compensated their family members (even children) through wages, shares of farm revenue, payment in kind, or other non-monetary means. However, most of the time, farmers and collectors did not have access to markets unless through intermediaries. Market places were not easily reachable by all the farmers and collectors. They preferred to sell their product to middlemen on the spot rather than travel to the town markets and try to sell a small amount of product. It was unusual for consumers and retailers to buy from them. Their main access to market was through their known middleman.

The intermediaries knew the market and consumer trends and set prices and conditions with the suppliers (growers or collectors). Market transparency was very weak and greatly restricted the market power of actors at the start of the chain (collectors of wild species and growers). Lack of market information was a common feature for NUS producers. Transparency of the market and better market information, particularly for the producers, should be created by market intelligence systems, improving the availability of data on demand and supply, production capacity and market prices, at least with regard to local markets.

There was a lack of trust and stability among chain actors, impeding the functioning of the market chains for the benefit of the rural communities. Cooperation did not exist and there were few efforts towards horizontal and vertical integration. To build trust, there is thus a need for cooperative effort among various market-chain actors and an enhancement of market literacy on the part of producers. Moreover, vertical integration of functions could help the producers to improve their market position by extending their involvement in the chain to include processing and trading activities. Improving trust among chain actors, fostering a more organized market structure, and investing in infrastructure related to the chain activities are all interventions that could be of great benefit for community livelihoods.

The collection of wild species identified as wild forest species was regulated by legislation that restricted both harvesting and trading. Calls for better regulatory systems have already been voiced by local community members and product manufacturers, who have realized that these regulations were hampering the species' full commercial potential, without endangering the conservation of the species. In addition, capacity building for agronomic practices (for sustainable harvesting from the wild and from cultivation), value-adding processing technologies, improved quality of storage, packaging and product diversification, should be provided for chain actors. The capacity building efforts, together with the reorganization of the market chain, could help facilitate the opening of small-scale processing factories or the development of local markets (and, in a few cases, provide access to international niche markets). Policy instruments should support the enhancement of market performance of NUS by supporting their sustainable conservation.

BIODIVERSITY AND ADAPTATION OF NUS

While crop production has become based on a limited number of plant species, the importance of the inter-species biodiversity of plant species has been shown by ethnobotanical surveys, indicating that hundreds of useful species—neglected and underutilized—were still being grown and managed in agricultural systems, particularly in developing countries, where they have demonstrated comparative advantages in terms of adaptability to low input agriculture and marginal lands (Padulosi et al. 2002). The NUS in this study are suitable for natural dryland environments: in most cases they are rain-fed or need little watering to be productive. Most of the species have their centre of origin in the Central and West Asia and North Africa region and are very adaptable to local climate conditions and soil, with good resistance to both abiotic and biotic stresses. They are tolerant of drought and salinity. They do not need special care and can grow without the aid of chemicals. However, the farmers that can afford pesticides and fertilizers (for other species that they grow) are using them to increase the productivity of the NUS. While growers cultivate NUS on private land on a very small scale, collectors gather the wild species either from communal or state lands. More than half of the collectors interviewed indicated that the quantity and the quality of the wild species that they were gathering had not changed compared with a decade previous. Since NUS are very adaptable to marginal environments, where the most fragile groups live, they represent a source of income of particular significance for those groups, in particular women and children, who can harvest these species from the wild, having land and labour access within the boundary of their community villages, often exploiting uncultivated areas. For this reason, the sustainable use of these species and their conservation is so important.

This pilot study showed that actors along the chain rarely recognized the existence of different species and varieties from a marketing point of view.

Generally, they referred to the most common variety as *baladi*, meaning 'local' in Arabic. Hence, the intra-varietal diversity has not yet been assessed and capitalized due to lack of specific knowledge about differences in final products resulting from differences in varieties and consumer appreciation. However, in the particular case of fig, the pilot study highlighted the positive correlation between cultivation of varieties and growers' income shares, and the negative correlation between fig varieties and traders' income shares. This might indicate that the market was focused on one or a few commercial varieties (probably better known, more appealing, etc.), while production was still based more on a number of varieties as a risk-reduction strategy to provide the grower with better income opportunities, as the harvest of specific varieties differed according to growing conditions in different years.

RECOMMENDATIONS

To prompt the small-scale farmer or rural household to grow, collect, process and trade these minor species, a range of benefits that can be perceived by the chain actors need to exist. These benefits can be identified as access to already existing and potential new local and international markets, or in the form of new products that can attract new markets at local and international levels. To reach these goals, a range of research and development implications have been identified and grouped at three levels: production, market and policy.

At production level, studies should be conducted to gather scientific information and provide genetic analyses of the species and their varieties so as to identify the best varieties and appropriate cultivation technology. Studies to assess the properties of different varieties and their adaptation to the environment should be carried out. Planting material should be provided to the farmers, with training on cultivation practices. To increase public awareness about the species and their uses, fact sheets on the species should be developed and distributed.

To overcome the constraints at market level, a range of recommendations for research and development have been identified. Stakeholder meetings, involving as many as possible of the market-chain actors involved in the NUS activities, including producers and traders, cultivation experts, non-profit organizations and representatives of relevant ministries, should be organized to verify that there are existing potentials and communities members committed to dedicating part of their time to further develop ideas for enhancing market potential. Such meetings would serve to create interest groups around one or more of the particular products, bringing all interested persons together. Each of these groups would have to be supported by an external facilitator from a research or development organization, whose job would be to ensure optimal interaction and mutual learning. In this participatory way (Annex 5), and with the help of marketing experts, new marketing options with a value-adding character should be researched within a frame of in-depth market studies of new products and new market access options, including 'fair-trade' channels. Through the application of such participatory market-chain analysis methodologies (for example, the participatory market chain approach—PMCA), trust among chain actors could be reinforced (Bernet et al. 2006).

Moreover, there is a need for identification of techniques for improving the processing of the products and of checking control mechanisms. Examples from experience in other arenas, and on other similar products, should be transferred, adapted and applied to the Syrian reality. This would involve establishing contacts with market-chain actors from other countries. This process would imply the construction of some infrastructure, such as processing, packaging and storage facilities, and facilitating access to and operation of simple processing machinery. This could be done through private partnerships, the involvement of public authorities, support from international organizations and private companies already operating in other countries. Training in value-addition processing technologies, enhanced storage techniques, improved packaging and product diversification should be offered to chain actors through farmer-to-farmer training, the extension services, product fairs, or rural theatre.

To overcome the problem of lack of access to markets, an in-depth study of the role of intermediaries should be conducted, and market options should be created to be directly linked to rural poor small-scale producers and processors. Some identified options are fair-trade channels, joint ventures with private sector entities in other countries, and the creation of local niche markets, such as tourism-oriented sales outlets. To analyse these options and to try to capitalize on underutilized species grown in remote areas using a marketing driven focus to conserve biodiversity and increase small-scale farmers' income, the Marketing Approach to Conserve Agricultural Biodiversity (MACAB) method could be applied (Annex 5). This method seeks to optimally involve the private sector in buying in to a previously developed product and elaborated marketing concept, to achieve best impact in terms of product image and market size (Bernet et al. 2003). There is need for government intervention in terms of capacity building, infrastructure and marketing intelligence.

At policy level, there should be development and adaptation of the existing legal framework, concerning:

- development of standards for processing and commercialization and relative certification (studies for the application of certification of geographical indications of origin);
- labelling regulation; and
- reformulation of regulations to support sustainable harvesting of wild species, which should not only safeguard the conservation of genetic resources, but also boost sustainable commercialization for the benefit of present and future generations of poor rural communities.

At conservation level, studies on the impact of current rates of harvest and collection of wild plants should be conducted. Sustainable gathering practices should be developed and transferred to collectors by training and the exten-

sion service. Research on cultivation practices should be carried out, and cultivated species should be grown close to wild species, to maintain gene flow and
continued evolution of these rustic species.

These recommendations can be summarized in a matrix, as in Table 29.

Table 29. *Matrix of recommendations.*

Production level	Characterization of traits, uses and adaptation
	Identification of best cultivation technologies
	Planting material and training on cultivation practices made available to farmers
	Development and distribution of information on the species and its use
Market level	Organization of meetings involving market-chain actors to discuss how to enhance market potential
	In-depth market studies on market options and market access (fair trade, joint ventures), using participatory analysis
	Exchange of knowledge with other market-chain actors in other countries
	Private and public partnerships for the construction of small infrastructure for the production of a better quality product
	Training of producers in improved processing techniques, better quality storage and packaging
Policy level	Reformulation of regulations aiming at environmental conservation, as well as maintaining the economic value of the wild species for poor rural communities
	Development and adaptation of the existing legal framework for quality standards and product labelling
Conservation level	Studies on the impact of unsustainable collection from the wild
	Training on sustainable gathering practices suitable for farmers and collectors
	Grow cultivated species close to wild species to maintain gene flow and continued evolution of these rustic species

POSITIVE NOTE
FOR FUTURE DEVELOPMENT

To allow these actions to take place, all stakeholders with an impact on market
development—from extensionists to policy-makers, from agricultural researchers to economists, to the private sector—should be aware of constraints and
opportunities for the development of markets, and consequently help build up
these local initiatives.

Bibliography

Abi-Antoun M, Chehabeddine H, Ojeil C. 2005. Current activities on conservation and sustainable use of underutilized plant resources. Paper presented at the *International Conference on Promoting Community-Driven Conservation and Sustainable Use of Dryland Agrobiodiversity*, ICARDA, Aleppo, Syria, 18–21 April 2005.

Albu M, Scott A. 2001. Understanding livelihoods that involve micro-enterprise: markets and technological capabilities in the SL framework. Intermediate Technology Development Group (ITDG), UK.

Al-Hakim W. 1994. Rapport sur les produits forestiers non-ligneux dans les forêts Syriennes (surtout dans le Département de Lattakia de Qunaitra). Forêts et Sécurité Alimentaire dans les Régions Méditerrannéenne et Moyen-Orientale. Field document of FAO project GCP/INT/539/ITA.

Al Ibrahim A. 1997. [Etude pomologique de sept varieties du fiquier (*Ficus carica* L.) typique de la region de Idleb.] 37eme Semaine Scientifique, Damascus, Syria. (in Arabic)

Alkire B. 1998. Capers. Purdue University New Crop FactSHEET. See: http://www.hort.purdue.edu/newcrop/cropfactsheets/caper.html

Arora RK, Ramanatha Rao V. (editors). 1998. Tropical Fruits in Asia: Diversity, Maintainance, Conservation and Use. Proceedings of the IPGRI-ICAR-UTFANET Regional Training Course on the *Conservation and Use of Germplasm of Tropical Fruits in Asia*, Indian Institute of Horticultural Research, Bangalore, India, 18–31 May 1997. IPGRI, Rome, Italy

Arroyo-García R, Martínez-Zapater JM, Fernández Prieto JA and Álvarez-Arbesú R. 2001. AFLP evaluation of genetic similarity among laurel populations (*Laurus* L.). *Euphytica* 122(1):155–164.

Barbera G, Lorenzo di R, Barone E. 1991. Observations on *Capparis* populations cultivated in Sicily and on their vegetative and productive behaviour. *Agricoltura Mediterranea* 121(1):32–39.

Bernet T, Hibon A, Bonierbale M, Hermann M. 2003. Marketing Approach to Conserve Agricultural Biodiversity. *In*: CIP-UPWARD. Conservation and Sustainable Use of Agricultural Biodiversity: A Sourcebook. User's perspectives with Agricultural Research and Development. CIP-UPWARD, Los Baños, Philippines. pp. 590–598.

Bernet T, Devaux A, Ortiz O, Thiele G. 2005. Participatory Market Chain Approach. *BeraterInnen News* 1/2005: 8–13. See http://www.lbl.ch/internat/services/publ/bn/bn-liste.htm

Bernet T, Thiele G, Zschocke T. (editors). 2006. Participatory Market Chain Approach (PMCA) – User Guide. CIP, Lima, Peru. http://papandina.cip.cgiar.org/fileadmin/PMCA/User-Guide.pdf

Best R, Ferris S, Schiavone A. 2005. Building linkages and enhancing trust between small-scale rural producers, buyers in growing markets and suppliers of critical inputs. *In* F.R. Almonds and S.D. Hainsworth (editors). Proceeding of the International Seminar: *Beyond Agriculture: Making Markets Work for the Poor*. Westminster, London, UK, 28 February-1 March 2005. Crop Post-Harvest

Programme (CPHP), Natural Resources International Ltd., Aylesford, UK, and Practical Action, Bourton on Dunsmore, UK. pp. 19–50. Available through the CPHP Web site: http://www.cphp.uk.com/

Biénabe E, Sautier D. 2005. The role of small scale producers' organizations in addressing market access. *In* F.R. Almonds and S.D. Hainsworth (editors). Proceeding of the International Seminar: *Beyond Agriculture: Making Markets Work for the Poor.* Westminster, London, UK, 28 February-1 March 2005. Crop Post-Harvest Programme (CPHP), Natural Resources International Ltd., Aylesford, UK, and Practical Action, Bourton on Dunsmore, UK. pp. 69–85. Available through the CPHP Web site: http://www.cphp.uk.com/

Blowfield M. 2001. Value chains. *Resource Centre for the Social Dimensions of Business Practice, Issue Paper* 2. 3 p. See: http://www.ethicaltrade.org/Z/lib/2001/other/nri-issuepa2.pdf

Buerli M, Aw-Hassan A. 2004. Microfinance in marginal dry areas: impact of village credits and savings associations on poverty in the Jabal al Hoss region in Syria. *In* K.J. Peters et al. (editors.) Proceedings of Deutscher Tropentag 2004 on *Rural Poverty Reduction through Research for Development and Transformation: international research on food security, natural resource management and rural development.* 5–7 October 2004, Berlin, Germany. Humboldt-Universität, Berlin, Germany.

Caillon S, Lanouguère-Bruneau V. 2003. Taro diversity in a village of Vanua Lava island (Vanuatu): where, what, who, how and why? Paper presented at Third Taro Symposium, Nadi, Fiji, 21–23 May 2003. See http://www.spc.int/cis/tarosym/TaroSym%20CD/index.htm

CBD [Convention on Biological Diversity]. 1992. http://www.biodiv.org/convention/convention.shtml

CBS [Central Bureau of Statistics]. 1999. Agricultural Census 1994. Office of the Prime Minister of the Syrian Arab Republic, Damascus.

CBS. 2001. Statistical Abstract 2001. Office of the Prime Minister of the Syrian Arab Republic, Damascus.

CGIAR [Consultative Group on International Agricultural Research]. 2004. Stakeholder Meeting. *In* Summary Record of Proceedings of the Annual General Meeting, Mexico City, Mexico, 27–28 October 2004.

Chweya JA, Eyzaguirre PB. (editors). 1999. *The biodiversity of traditional leafy vegetables.* IPGRI, Rome, Italy. 182 p.

de Groot P, Haq N. (editors). 1995. *Promotion of Traditional and Underutilized Crops.* Report of a workshop held in Valletta, Malta, June 1992. [Commonwealth Science Council] Series No. CSC(95)AGR23 Technical Paper 311. ICUC/CSC, London, UK.

DFID [Department for International Development]. 1999. Sustainable Livelihoods Approach Guidance Sheets. DFID, London, UK. http://www.livelihoods.org/info/info_guidancesheets.html#6

Dorward A, Poole N, Morrison J, Kydd J, Urey I. 2002. Critical Linkages: Livelihoods, Markets and Institutions. Paper presented at the Seminar on *Supporting Institutions, Evolving Livelihoods.* Bradford Centre for International Development, University of Bradford, UK, 29–30 May 2002.

Ellis, F. 2000. Rural Livelihoods and Diversity in Developing Countries. Oxford University Press, Oxford, UK.

Engels JMM, Ramanatha Rao V, Brown AHD, Jackson MT. (editors). 2002. *Managing Plant Genetic Diversity.* CABI, Wallingford, UK, and IPGRI, Rome, Italy. 487 p.

Eyzaguirre P, Padulosi S, Hodgkin T. 1999. IPGRI's strategy for neglected and underutilized species and the human dimension of agrobiodiversity. *In* S. Padulosi (editor). *Priority-setting for underutilized and neglected plant species of the Mediterranean region.* Report of the IPGRI Conference, ICARDA, Aleppo. Syria, 9-11 February 1998. IPGRI, Rome, Italy. pp. 1–20.

FAO [Food and Agriculture Organization of the United Nations]. 1995. Non-wood forest products in nutrition. Paper prepared for the FAO/GOI Expert Consultation on Non-Wood Forest Products, Yogyakarta, Indonesia, 17–27 January 1995.

FAO. 1996. Global Plan of Action for the Conservation and Sustainable Utilization of Plant Genetic Resources for Food and Agriculture, and Leipzig declaration, adopted by the International Technical Conference on Plant Genetic Resources, Leipzig, Germany, 17–23 June 1996. FAO, Rome, Italy.

FAO. 2000. Community-based tree and forest product enterprises: Market Analysis and Development Field Manual. Prepared by I. Lecup and K. Nicholson. FAO, Rome.

FAO. 2003. Market research for agroprocessors. Prepared by A.W. Shepherd. *[FAO] Marketing Extension Guide*, no. 3. See http://www.fao.org/waicent/faoinfo/agricult/ags/AGSM/markres.pdf

GEF [Global Environment Facility]/UNDP and Syrian Arabian Republic Ministry of Environment (National Biodiversity Unit [NBU]). 2002. Biological Diversity National Report, Biodiversity Strategy and Action Plan and Report to the Conference of the Parties. NBSAP Project SY/97/G31

Gereffi G. 1994. The organization of buyer-driven global commodity chains: How U.S. retailers shape overseas production networks. *In* G. Gereffi and M. Korzeniewicz (editors). *Commodity Chains and Global Capitalism*. Praeger, London, UK.

GFU [Global Facilitation Unit for Underutilized Species]. 2002. Conceptual framework of the multi-stakeholder initiative established under the umbrella of the Global Forum on Agricultural Research (GFAR) and hosted by the International Plant Genetic Resources Institute (IPGRI). See: www.underutilized-species.org

Griffon M. (coordinator). 2002. Filières agroalimentaires en Afrique: comment rendre le marché plus efficace? DGCID, Série rapports d'étude, Ministère des Affaires Etrangères, 314 p. (étude de cas de G. Duteurtre sur la filière laitière périurbaine de Niono, Mali). See: http://www.france.diplomatie.fr/cooperation/dgcid/publications/etudes_01/ agroalimentaires/pdf/doc26.pdf

Guendel S, Hoeschle-Zeledon I, Krause B, Probst K. (editors). 2003. Proceedings of the International Workshop on *Underutilized Plant Species and Poverty Alleviation*. Leipzig, Germany, 6–8 May 2003. See: http://www.underutilized-species.org/the_latest/archive/pop_up/leipzig%20proceedings.pdf

Hellin J, Griffith A, Albu M. 2005. Mapping the market: market literacy for agricultural research and policy to tackle rural poverty in Africa. *In* F.R. Almonds and S.D. Hainsworth (editors). Proceeding of the International Seminar: *Beyond Agriculture: Making Markets Work for the Poor*. Westminster, London, UK, 28 February-1 March 2005. Crop Post-Harvest Programme (CPHP), Natural Resources International Ltd., Aylesford, UK, and Practical Action, Bourton on Dunsmore, UK. pp. 19–50. Available through the CPHP Web site: http://www.cphp.uk.com/http://www.cphp.uk.com/uploads/documents/CPHP%20Overview%20Paper%20Final%20Version.pdf

Hellin J, Higman S. 2005. Crop Diversity and Livelihood Security in the Andes. *Development in Practice*, 15(2): 165–174.

Hillman G. 1975. The plant remains from Tell Abu Jureyra. *Proceedings of the Prehistory Society* 41: 70–73.

Hobley M. 2001. Unpacking the PIP box. Discussion paper on www.livelihoods.org. Hobley Shields Associates, Chard, Somerset, UK.

Hugon, P. 1994. Filières agricoles et politiques macroéconomiques en Afrique subsaharienne. *In* M. Benoit-Cattin, M. Griffon and P. Guillaumont. (editors). *Economie des politiques agricoles dans les pays en développement*. Revue Française d'Economie, Paris, France.

ICARDA [International Centre for Agricultural Research in the Dry Areas]. 2000. Promotion of *in situ* conservation of wild fruit trees: example of *Ziziphus* species at Al-Haffeh, Syria. *Dryland Agrobio* No. 2: 8. (July–September 2000 newsletter of the GEF/UNDP Project on Conservation and Sustainable Use of Dryland Agrobiodiversity). See: http://www.icarda.org/Gef/NewsLetter28.HTML

ICARDA. 2001. Figs in Syria. *Dryland Agrobio* No.4: 1–2. (January–March 2001 newsletter of the GEF/UNDP Project on Conservation and Sustainable Use of Dryland Agrobiodiversity). See: http://www.icarda.org/Gef/Agro4.pdf

ICUC [International Centre for Underutilised Crops]. 2001. Fruits for the future - Ber. Factsheet No. 2. February 2001, Southampton, UK. See: http://www.icuc-iwmi.org/files/Resources/Factsheets/ziziphus.pdf

ICUC. 2002. Ber Extension Manual. ICUC, Southampton, UK.

IPGRI [International Plant Genetic Resources Institute]. 2002. Neglected and Underutilized Plant Species: Strategic Action Plan of the International Plant Genetic Resources Institute. IPGRI, Rome, Italy. Available from: http://www.ipgri.cgiar.org/index.htm

IPGRI and CIHEAM. 2003. Descriptors for Fig. International Plant Genetic Resources Institute, Rome, Italy, and International Centre for Advanced Mediterranean Agronomic Studies, Paris, France. See: http://www.ipgri.cgiar.org/publications/pdf/907.pdf

IPGRI/GFU/MSSRF [MS Swaminathan Research Foundation]. 2005. Meeting the Millennium Development Goals with Agricultural Biodiversity. IPGRI, Rome, Italy.

Johns T, Eyzaguirre PB. 2002. Nutrition and the environment. *In* Nutrition: A Foundation for Development. (Brief 5 of 12). IFPRI and UN ACC/SCN, Geneva, Switzerland. See: http://www.ifpri.org/

Johns T, Sthapit BR. 2004. Biocultural diversity in the sustainability of developing country food systems. *Food and Nutrition Bulletin* 25(2): 143–155. Available at: http://www.ipgri.cgiar.org/themes/human/publications/FoodNutritionBulletinJohns.pdf

Kaplinsky R, Morris M. 2001. A Handbook for Value Chain Research. Prepared for IDRC. No publisher. See: www.ids.ac.uk/ids/global/pdfs/VchNov01.pdf

Khouildi S, Pagnotta MA, Tanzarella OA, Ghorbel A, Porceddu E. 2000. Suitability of RAPD (random amplified polymorphic DNA) technique for estimating the genetic variation in natural genotypes of Tunisian and Italian caper (*Capparis spinosa* L.). *Agricoltura Mediterranea* 130(1):72–77.

Kydd J. 2002. Agriculture and rural livelihoods: is globalization opening or blocking paths out of rural poverty? Agricultural Research and Extension Network Paper No. 121. ODI, London, UK.

Lawrence, GHM. 1951. *Taxonomy of Vascular Plants*. MacMillan Company, New York, USA.

Lundy M, Gottret MV, Cifuentes W, Ostertag CF, Best R, Peters D, Ferris S. 2004. Increasing the Competitiveness of Market Chains with Smallholder Producers. Field Manual 3. The Territorial Approach to Rural Agro-enterprise Development. CIAT, Cali, Colombia. 117 p. Available from: http://www.ciat.cgiar.org/agroempresas/pdf/manual3_marketchain.pdf

MAAR [Ministry of Agriculture and Agrarian Reform of the Syrian Arab Republic]. 2002. The Annual Agricultural Statistical Abstract, 2002.

MAAR-NAPC [National Agricultural Policy Centre of the Syrian Arab Republic]. 2003. Syrian Agriculture Database 2003. Activity under FAO Project GCP/SYR/006/ITA - Phase II *Assistance for Capacity Building through Enhancing Operation of the National Agricultural Policy Centre*. FAO-Government of Italy Cooperative Programme.

Magness JR, Markle GM, Compton CC. 1971. Food and feed crops of the United States. A descriptive list classified according to potentials for pesticide residues. Interregional Research Project IR-4, Bulletin 1 (Published as New Jersey Agricultural Experiment Station Bulletin No. 828.)

Mars M. 2003. Fig (*Ficus carica* L.). Genetic Reources and Breeding. II International Symposium on Fig. *ISHS Acta Horticulturae* 605:19–27. See: http://www.actahort.org/books/605/605_1.htm

McClintock NC. 2004. Roselle in Senegal and Mali. *LEISA Magazine* 20(1): 8–10. See: http://www.leisa.info/

Mengjun L. 2003. Genetic diversity of Chinese jujube (*Ziziphus jujuba* Mill.). In: XXVI International Horticultural Congress: Plant Genetic Resources, The Fabric of Horticulture's Future. *ISHS Acta Horticulturae* 623:351–355.

Miller Cavitch S. 1997. The Soapmaker's Companion: A Comprehensive Guide with Recipes, Techniques & Know-How. Storey Books, North Adams, MA, USA.

Monti L. (editor). 1997. Proceedings of the CNR International Workshop on *Neglected Plant Genetic Resources with a Landscape and Cultural Importance for the Mediterranean Region.* Naples, Italy, 7–9 November 1996. Consiglio Nazionale delle Ricerche. SMED-CNR Project. Naples, Italy.

MSSRF [MS Swaminathan Research Foundation]. 1999. *Enlarging the basis of food security: the role of underutilized species.* Proceedings of the international consultation organized by the Genetic Resources Policy Committee of the CGIAR, held at the M.S. Swaminathan Research Foundation (MSSRF), Chennai, India, 17-19 February 1999.

NAPC [National Agricultural Policy Centre]. 2003. The State of Food and Agriculture in the Syrian Arab Republic in 2002. With the support of FAO Project GCP/SYR/006/ITA. See: http://www.napcsyr.org/dwnld-files/periodical_reports/SOFAS%202002%20En.pdf

New Agriculturist. 2004. Focus on... Underutilised crops. Reporting agriculture for the 21st century. New Agriculturist-on line. June 2004. See: http://www.new-agri.co.uk/04-6/focuson.html

Padulosi S, Hodgkin T, Williams JT, Haq N. 2002. Underutilized crops: trends, challenges and opportunities in the 21st century. *In* JMM Engels, V Ramanatha Rao, AHD Brown and MT Jackson (editors). *Managing Plant Genetic Diversity.* CABI, Wallingford, UK, and IPGRI, Rome, Italy. pp. 323–338. See: http://www.ipgri.cgiar.org/Publications/727/pdf/0851995225Ch30.PDF

Padulosi S, Noun J, Giuliani A, Shuman F, Rojas W, Ravi B. 2003. Realizing the benefits in neglected and underutilized plant species through technology transfer and Human Resources Development. *In* Proceeding of the *Norway/UN Conference on Technology Transfer and Capacity Building,* 23–27 May, 2003, Trondheim, Norway. pp. 117–127. See: http://www.underutilized-species.org/documents/trondheim_paper_03.pdf

Padulosi S, Hoeschle-Zeledon I. 2004. Underutilized plant species: what are they? *LEISA Magazine* 20(1): 5–6. See: http://www.leisa.info/FritZ/source/getblob.php?o_id=65172&a_id=211&a_seq=0

Patton M. 2002. *Qualitative Research and Evaluation Methods.* 3rd edition. Sage Publications.

Porter M. 1990. The Competitive Advantage of Nations. Macmillan, London, UK.

Prescott-Allen R, Prescott-Allen C. 1990. How many plants feed the world? *Conservation Biology* 4: 365–374.

Rivera D, Inocencio C, Obon C, Alcaraz F. 2003. Review of food and medicinal uses of *Capparis* L. subgenus *Capparis* (Capparidaceae). *Economic Botany* 57(4): 515–534.

Sarris A. 2001. Agricultural Development Strategy for Syria. Document prepared under FAO Project GCP/SYR/006/ITA. See: http://www.napcsyr.org/dwnld-files/policy_studies/en/16b_agr_dev_strategy_Syria_en.pdf

Sarris A. 2003. Agriculture in the Syrian macroeconomic context. *In* C. Fiorillo and J. Vercueil (technical editors). Syrian agriculture at the crossroads. Prepared under Project GCP/SYR/006/ITA *–Assistance for Capacity Building through Enhancing Operation of the National Agricultural Policy Centre. FAO Agricultural Policy and Economic Development Series,* No. 8. See: http://www.napcsyr.org/sac.htm

SEBC [Syrian European Business Centre]. 2003. Ghar Soap Catalogue.

Simon JE, Chadwick AF, Craker LE. 1984. *Herbs: An Indexed Bibliography. 1971-1980.* The Scientific Literature on Selected Herbs, and Aromatic and Medicinal Plants of the Temperate Zone. Archon Books, Hamden, CT, USA. 770 p.

Simopoulos AP, Norman HA, Gillaspy JE, Duke JA. 1992. Common purslane: a source of omega-3 fatty acids and antioxidants. *Journal of the American College of Nutrition* 11(4): 374–382.

Smale M, Bellon MR. 1999. A conceptual framework for valuing on-farm genetic resources. *In:* D. Wood and J.M. Lenné (editors). *Agrobiodiversity: Characterization, Utilization and Management.* CABI Publishing, Wallingford, UK. pp. 387–408.

Smale M, Bellon M, Jarvis D, Sthapit B. 2004. Economic concepts for designing policies to conserve crop genetic resources on farms. *Genetic Resources and Crop Evolution* 51: 121–135.

Tous J, Ferguson L. 1996. Mediterranean fruits. *In:* J. Janick (editor). *Progress in New Crops.* ASHS Press, Arlington, VA, USA. pp. 416–430.

Trochim WM. 2001. The Research Methods Knowledge Base. 2nd edition. Cornell University, USA.

von Maydell HJ. 1989. Criteria for the selection of food-producing trees and shrubs in semi-arid regions. *In* G.E. Wickens, N. Haq and P. Day (editors). *New Crops for Food and Industry.* Chapman and Hall, London, UK. pp. 66–75.

UNCTAD. 2000. The Post-Uruguay Round tariff environment for developing country exports: tariff peaks and tariff escalations. UNCTAD/WTO Joint Study. Doc. No. TD/B/COM.1/14/Rev. 1, 28 January 2000. See: http://www.unctad.org/Templates/Download.asp?docID=333&intItemID=2068&lang=1

WIEWS [World Information and Early Warning System on PGRFA]. 2002. Global survey 2002. FAO, Rome, Italy. See: http://apps3.fao.org/wiews

Annex 1
The sustainable
livelihoods framework

Human asset represents the skills, knowledge, ability to labour and good health that together enable people to pursue different strategies and achieve their livelihood objectives. At a household level, human capital is a factor of the amount and quality of labour available, and varies according to household size, skill levels, leadership potential, health status, etc. Some human capital indicators are household size, health status and health service access, living standards, education level, training access for gender, labour experience and skills. Human capital is a necessary pillar required in order to take advantage of any of the four other types of capital.

Social asset is a second type of asset. Inside the livelihood approach, social capital is taken as all the social resources employed by individuals to seek their livelihood objectives, through networks and connectedness (vertical or horizontal), membership of more formalized groups, and relationships of trust, reciprocity and exchange, all interrelated. Social capital is important as a 're-source of last resort' for the poor and vulnerable, providing a buffer that helps them cope with shocks, acting as an informal safety net to ensure survival, and compensating for lack of other types of capital. At the same time, it is directly linked with financial capital (by improving the efficiency of economic relations, it can help increase people's incomes and rates of saving), natural capital (it can be effective in improving the management of common resources), physical capital (it can be effective in the maintenance of shared infrastructure), and there is even a close relationship with human capital (social networks facilitate innovation, the development of knowledge and sharing of that knowledge).

Natural asset is formed by all the natural resources, from intangible public goods (atmosphere, biodiversity, etc.) to divisible assets used directly for production (trees, land, etc.). By its definition, natural capital is sensitive to the external environment (e.g. fires that destroy forests; floods and earthquakes that destroy agricultural land).

Natural capital is crucial to people who base their livelihoods (in whole or in part) on natural-resource-based activities (farming, fishing, wild collection from forests, mineral extraction, etc.). It is also fundamental for health and well-being, showing a clear linkage with human capital. Examples of natural capital and services deriving from it (DFID 1999) include land, forests, marine resources, wild resources, water, air quality, ecosystem protection, waste assimilation, storm protection, and biodiversity.

For all these it is important to consider access and quality, and how both are changing.

Financial asset indicates the financial resources necessary to realize the people's livelihood objectives. The two main sources of financial capital are the *available stocks*—savings in cash, deposit, livestock and jewellery, as well as credits—and *regular inflows of money*—all types of inflows, excluding earned income. The principal characteristic of financial capital is the versatility (financial capital can be converted into other types of capital; be used for direct achievement of livelihood outcome; be transformed into political and decision-making process influence; and govern access to resources [DFID 1999]), but the financial resources are often misused due to lack of knowledge and constrained by policies with negative effects on underdeveloped markets and small-scale enterprises. Furthermore, financial capital is the least available resource to the poor. This is why the other asset categories are so important to them.

Physical asset includes all the basic infrastructure and producer goods (tools and equipment used to function more productively), needed to support livelihoods. Examples of essential components of infrastructure are (DFID 1999): affordable transport; secure shelter and buildings; adequate water supply and sanitation; clean, affordable energy; and access to information (communications). The opportunity costs associated with poor infrastructure can preclude education, access to health services and income generation. Moreover, scarce and inadequate producer goods also constrain people's productive capacity, and consequently their human capital. Need for physical capital must be perceived and required from the users in order to maintain services in a sustainable manner.

The livelihood framework embeds various aspects interlinked with livelihood assets: the vulnerability context; Policies, Institutions and Processes (PIP); and Livelihood Strategies and outcomes (Figure A1.1).

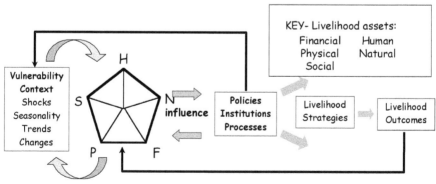

Source: DFID 1999

Figure A1.1 *The Sustainable Livelihoods Framework.*

Annex 2
Household field
survey questionnaires

COLLECTORS' QUESTIONNAIRE

- Household composition
- Household wealth
- Labour data
- **Environmental conditions in which the wild species grow**
- **Collection data**
- Data related to reason for carrying out the activities and the output
- Data related to legal and policy issues
- Marketing Data

GROWERS' QUESTIONNAIRE

- Household composition
- Household wealth
- Labour data
- **Cultivation Data**
- **Collection Data**
- Data related to reason for carrying out the activities and the output
- Data related to legal and policy issues
- Marketing Data

PROCESSORS' QUESTIONNAIRE

- Household composition
- Household wealth
- Labour data
- **Data related to transformation activity**
- Data related to reason for carrying out the activities and the output
- Data related to legal and policy issues
- Marketing Data

COLLECTORS' QUESTIONNAIRE

- Household composition
- Household wealth
- Labour data
- **Data related to trading activity**
- Data related to reason for carrying out the activities and the output
- Data related to legal and policy issues
- Marketing Data

Annex 3
Agro-climatic zones and agricultural regions in Syria

Five agro-climatic zones are identified on the basis of rainfall patterns (Figure A3.1).

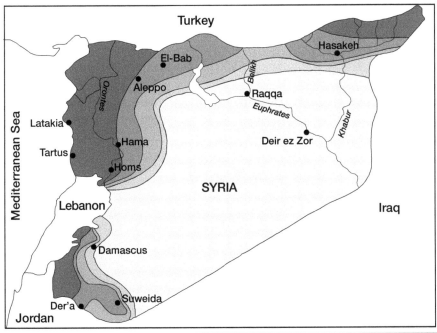

		Annual Precipitation (mm)	Bioclimatic Region	Ratio of land (%)
	Zone 1	>350	Humid, semi-humid, arid	14.6
	Zone 2	>250–350 (>250, >2/3 year)	Arid	13.4
	Zone 3	>250 (>250, >1/2 year)	Arid	7.1
	Zone 4	200–250 (>200, >1/2 year)	Arid	9.8
	Zone 5	<200	Very-arid	55.1

Source: MAAR 2002

Figure A3.1 *Agro-climatic zones in Syria.*

ZONE 1

Zone 1 is 2 701 000 ha, 14.6% of the national territory. The annual average rainfall is >350 mm. It is divided into two areas:
- Areas with an annual average rainfall of >600 mm, where rain-fed crops can be successfully planted; and
- Areas with an annual average rainfall of between 350 and 600 mm, but >300 mm during two-thirds of the monitored years, and where it is possible to get two successful crop seasons every three years. The main crops are wheat, legumes and summer crops.

ZONE 2

Zone 2 is 2 475 000 ha, 13.3% of the country. The average annual rainfall is 250–350 mm, but >250 mm during two-thirds of the monitored years; it is possible to get two barley crops every three years. Beside barley, wheat, legumes and summer crops are grown.

ZONE 3

Zone 3 is 1 830 000 ha, 7.1% of the country. Average annual rainfall is 250–350 mm, with >250 mm during half of the monitored years. The main crop is barley, but legumes could be planted.

ZONE 4

Zone 4 is 1 830 000 ha, 9.9% of the country, and is the marginal zone between the arable zones and the desert, with average annual rainfall of 200–250 mm, with >200 mm during half of the monitored years. This zone is suitable only for barley or for permanent grazing.

ZONE 5

Zone 5 is 10 209 000 ha, 55.1% of the country. It is desert and steppe, and it is not suitable for rain-fed cropping.

SYRIAN AGRICULTURAL REGIONS

Syria is divided into five main agricultural regions, namely Southern, Central, Coastal, Northern, and Eastern (NAPC 2003).

The Southern region (ca 15.7% of the country) includes Damascus, Dar'a, Al Sweida and Quneitra. It is famous for its fruit production, especially apricots, apples and grapes, although it also produces chickpea, tomato and cattle. In 1998–99, the regional contribution to national production was 35.8% for chickpea, 50.7% for apple, 31.2% for grape and 62.5% for apricot.

The Central region accounts for ca 27.6% of the total area, and produces mainly sugar beet, onion, potato and almond. In 1998–99, the regional contribution to national production was 57.2% for sugar beet, 52.6% for onion (dried), 31.3% for potato and 14.3% for irrigated wheat.

The Coastal region along the Mediterranean seaboard includes the cities of Lattakia and Tartous. Although this region is relatively small, it makes a significant contribution to national agricultural production, with 98% of citrus, 42% of olive, 55% of tomato and 56% of tobacco in 1998–99.

The Northern region covers 12.6% of the country and includes the cities of Aleppo and Idleb. Its main products are lentil (55% of national output), chickpea (51%), olive (56%) and pistachio (69%). Local farmers raise about 20% of the all sheep in Syria.

The Eastern region is the largest, concentrating national cereal and cotton production. In order to enhance productivity through irrigation, many networks have been built in this region, especially along the Euphrates and Al Khabour rivers, in addition to the many wells that have been excavated. Farms tend to specialize in irrigated wheat (64% of national production), rain-fed wheat (38%), cotton (63%) or lentil (29%).

Annex 4
SWOT analysis
of a potential joint venture
for caper production and trading

During the survey, there was the opportunity to discuss with a Syrian trader the possibility of establishing a caper market in Syria, through a joint venture with an Italian caper producer with a stable market. The same discussion was conducted with the manager of the Italian company. Table A4.1 shows a quick appraisal of the advantages and disadvantages of a potential joint venture between the Italian cooperative and a Syrian private partnership in terms of improving the market for capers. The appraisal was made on the basis of discussion with a Syrian businessman and a visit to the manager of the Italian cooperative of caper producers on the Island of Pantelleria, Italy. The Italian cooperative might be interested in planting their cultivar in Syria, where the capers could be cultivated by Syrian local farmers and processed according to Italian standards. The cooperative would buy the capers and package and distribute them in Italy with a new trademark. Assuming that there would be an intermediary collecting capers from them, the Syrian farmers interviewed in the remote area of Jabal-Al-Hoss, south-east of Aleppo, would be very willing to cultivate the capers, collect them and to learn how to process them to meet market demands.

Table A4.1 *SWOT (strengths – weaknesses – opportunities – threats) analysis of a potential partnership between Syrian businessmen and an Italian caper cooperative.*

Strengths	Weaknesses	Opportunities	Threats
Promotion of cultivation and commercialization of capers in Syria.	Business orientation: maximum profit provided to businessmen, with no immediate benefit for poor small-scale farmer livelihoods.	Opportunities for poor small-scale farmers, in terms of more employment in cultivated caper plantings, and maybe possibility to adopt this activity for themselves.	Introduction of an external cultivar (variety) could bring erosion of the local variety, and threaten the ecosystem. No improvement in labour conditions if the farmers growing capers for the businessmen do not have contractual power.

Annex 5
The Participatory Market Chain
Approach and MACAB

The Participatory Market Chain Approach (PMCA) is a new participatory R&D method recently developed by the International Potato Center (CIP), Peru. Involving various market-chain actors and supporting R&D organizations, it seeks to generate group innovations based on a well-led and well-structured participatory process that aims to stimulate interest, trust and collaboration among members of the chain. The innovations can be new products and processes, new technologies or new institutions, benefiting the actors directly or indirectly. The approach foresees three phases, of flexible duration (Figure A5.1), depending on the context of application (Bernet et al. 2005).

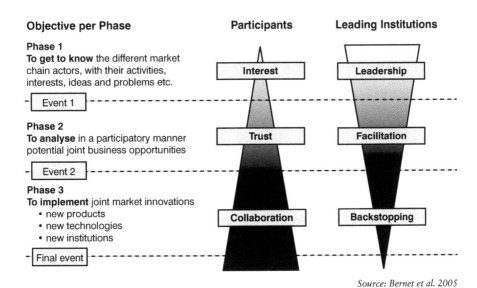

Source: Bernet et al. 2005

Figure A5.1 *The Participatory Market Chain Approach (PMCA) – objectives and structure.*

This approach was applied within the Papa Andina project, coordinated by CIP in Peru, aimed at identifying markets for native potato varieties (i.e. landraces), where small-scale farmers have a long-term competitive advantage because of their location, local knowledge, access to local varieties and special crop management practices. The project facilitates contacts between small-scale potato farmers and the processing companies that are becoming increasingly important buyers of potatoes. Farmers learn more about the processors' demands in terms of preferred potato varieties, volumes, quality and timing of production. The processors learn about the varieties of potatoes grown by the farmers, their quality aspects and how they are influenced by different growing schemes. With a greater understanding of the existing reality of producers and the established contacts, the processors can exploit potato varieties that were not used before (Hellin et al. 2005).

The Marketing Approach to Conserve Agricultural Biodiversity (MACAB) is designed to target and exploit new market opportunities that enhance both farmers' income and *in situ* agricultural biodiversity. The method consists basically of nine steps, defining a pathway from discovering interesting product attributes to product launching (Table A5.1).

Table A5.1 *The nine steps of the marketing approach to conserve agricultural biodiversity (MACAB) process.*

MACAB step	Key idea of each step
1. Discovery of promising crop attributes	To find "whys" for potential consumption and expansion of the crop
2. Development of a potential new product	To determine a way of consumption that is attractive and convenient for consumers
3. Analysis of the economic feasibility of the product	To make sure that production costs are not too high, thus making the product competitive
4. Elaboration of a sound marketing concept	To define optimal packaging and pricing to reach target consumers
5. Testing of the marketing concept with consumers	To fine-tune the concept and to measure real purchase interest and market size
6. Protection of brand name and concept	To prevent misappropriation and misuse of the marketing concept by entrepreneurs
7. Definition of criteria to select private enterprises	To justify the selection of enterprises that offer the highest likelihood of social impact
8. Transparent transfer of the 'marketing package' to private enterprises	To hand the business opportunity over to private enterprises (e.g. with a contract that enables the use of the brand under specified conditions)
9. Examination of enterprise behaviour and social impact	To make sure that the enterprise reaches the target social impact

Source: Bernet et al. 2003.

CIP has been applying MACAB in two crops: native potatoes and yacon (*Polymnia sonchifolia* Poeppig & Endl.). Promising product attributes of both crops were the starting point of the R&D process. At a later stage marketing specialists were hired to elaborate a concept to improve the image and use of these underutilized crops (Bernet et al. 2003)

INDEX